# Memory Based Question Bank on Horticulture

NIPA® GENX ELECTRONIC RESOURCES & SOLUTIONS P. LTD.
New Delhi-110 034

# Memory Based Question Bank on Horticulture

**Niranjan Singh**
Assistant Professor
Department of Fruit Science
Acharya Narendra Deva University of Agriculture and Technology
Ayodhya (UP), India

**D.P. Sharma**
Professor and Head
Department of Fruit Science
Dr. Yashwant Singh Parmar University of Horticulture & Forestry
Nauni, Solan, Himachal Pradesh

**Neeraj Sankhyan**
Assistant Professor
Dr. Yashwant Singh Parmar University of Horticulture & Forestry
Nauni, Solan, Himachal Pradesh

**NIPA® GENX ELECTRONIC RESOURCES & SOLUTIONS P. LTD.**
New Delhi-110 034

**NIPA® GENX ELECTRONIC**
**RESOURCES & SOLUTIONS P. LTD.**

101,103, Vikas Surya Plaza, CU Block
L.S.C.Market, Pitam Pura, New Delhi-110 034
Ph : +91 11 27341616, 27341717, 27341718
E-mail:newindiapublishingagency@gmail.com
www: www.nipabooks.com

***For customer assistance, please contact***
Phone: + 91-11-27 34 17 17
Fax: + 91-11-27 34 16 16
E-Mail: feedbacks@nipabooks.com



ISBN: 978-81-19235-69-8



Composed and Designed by NIPA.

# Preface

Memory Based Question Bank On Horticulture as the name suggest is book with questions based on previous examinations covering all the subjects of horticulture division.

This publication is beneficial for JRF, SRF, NET, ARS, M.Sc. Entrance, State, and Central Level Competitive Examinations. It is designed to be an objective-oriented and user-friendly book.

**Authors**

# Contents

*Preface* ........ *v*

1. ICAR – JRF Horticulture Exam – 2010 ........ 1
2. ICAR – JRF Horticulture Exam – 2011 ........ 19
3. ICAR – JRF Horticulture Exam – 2012 ........ 37
4. ICAR – JRF Horticulture Exam – 2013 ........ 55
5. ICAR – JRF Horticulture Exam – 2014 ........ 73
6. ICAR – JRF Horticulture Exam – 2015 ........ 91
7. ICAR – JRF Horticulture Exam – 2016 ........ 109
8. ICAR – JRF Horticulture Exam – 2022 ........ 127
9. ICAR – JRF Horticulture Exam – 2023 ........ 153
10. ICAR – SRF Horticulture Exam – 2015 ........ 179
11. ICAR – SRF Horticulture Exam – 2016 ........ 201

# 1

# ICAR – JRF Horticulture Exam – 2010

1. Horticulture sector covers % of total cropped area.
   a) 10% b) 15%
   c) 20% d) 25%
2. Which of the following is / are pre-emergence herbicide(s)?
   a) Diquat b) Paraquat
   c) Diuron d) All of these
3. Potassium plays an important role in:
   a) Osmotic regulation b) Opening and closing of stomata
   c) Sugar translocation d) All of these
4. Which of the following is a variety of jackfruit?
   a) Halawi b) Khadrawi
   c) Rudhrakshi d) Banarsasi Kadaka
5. Root and shoot pruning is a very common practice in:
   a) Rose b) Guava
   c) Jamun d) Gladiolus
6. 'Arka Pragati' variety of onion is suitable for cultivation in:
   a) Kharif season b) Rabi season
   c) Zaid season d) Both kharif and rabi
7. The shape of 'Scarlet Globe' variety of radish is:
   a) Cylindrical b) Round
   c) Square d) None of these
8. The tropical vegetable which is grown in the north-eastern region of India is:
   a) Coleus b) Elephant foot yam
   c) Topioca d) Sweet potato

9. Blanket flower is the common name for:
   a) Gaillardia  b) Gomphrena
   c) Amaranthus  d) Zinnia

10. Which of the following avenue tree is suitable for growing in saline and alkaline soil?
   a) *Tamarix gallica*  b) *Ficus religiosa*
   c) *Salix babylonica*  d) *Grevillea robusta*

11. Medicinal use of guggul (*Commiphora wightii*) is to:
   a) Reduce cholesterol  b) Reduce blood pressure
   c) Reduce tension  d) Reduce arthritis

12. BER, a physiological disorder of tomato is caused due to the deficiency of:
   a) Ca  b) B
   c) K  d) Mg

13. Which of the following rootstock of grape is suitable for saline soil?
   a) Dogridge  b) St. George
   c) Riparia Glory  d) Temple

14. The medicinal plant having alkaloids and edible nutritional seeds are:
   a) Periwinkle  b) Sarpagandha
   c) Pipali  d) Opium poppy

15. Mangala is a variety of:
   a) Arecanut  b) Coconut
   c) Cashew nut  d) Pecan nut

16. Citrus is highly susceptible to injury:
   a) Cl  b) Na
   c) CaCO3  d) All of these

17. Which of the following is an indigenous fruit crop?
   a) Guava  b) Papaya
   c) Loquat  d) Wood apple

18. The best intercrop planted in the banana orchard is:
   a) Papaya  b) Pineapple
   c) Ginger  d) Pepper

19. Which of the following is a form of repair grafting?
    a) Bridge grafting b) Tongue grafting
    c) Cleft grafting d) Wedge grafting
20. Which of the following fruit has non-endospermic (exalbuminous) seed?
    a) Mango b) Ber
    c) Grapes d) Papaya
21. Delphinidin, a plant pigment, is responsible for the development of ............ colour:
    a) Red b) Blue
    c) Pink d) Purple
22. Botanical name of cauliflower is:
    a) *Brassica oleracea* var. capitata b) *Brassica oleracea* var. gemmifera
    c) *Brassica oleracea* var. botrytis d) *Brassica oleracea* var. italica
23. Which of the following is the progenitor of cucumber?
    a) *Cucumis melo* b) *Cucumis anguria*
    c) *Cucumis lanatus* d) *Cucumis sativus* var. hardwickii
24. The ETL of aphid population in case of potato is:
    a) 20/100 leaves b) 30/100 leaves
    c) 40/100 leaves d) 50/100 leaves
25. Clones are generally degenerate due to:
    a) Viral infection b) Mechanical mixture
    c) Mutations d) All of these
26. Which of the following amino acid is the precursor of ethylene biosynthesis?
    a) Cysteine b) Proline
    c) Methionine d) Purine
27. The term 'Bareja system' is related to:
    a) Nutmeg b) Allspice
    c) Betelvine d) Betel nut
28. The term heterosis was coined by:
    a) East b) Shull
    c) Davenport d) East and Shull

29. Generally, the frequency of spontaneous mutation is:
    a) 10-6 b) 10-5
    c) 10-4 d) 10-7
30. Pusa Mukta is a black rot-resistant variety of:
    a) Cabbage b) Cauliflower
    c) Knol khol d) Brussels sprouts
31. 'Sui generis" is a Latin phrase which means:
    a) A new system b) A secure system
    c) A system offshoot d) A conventional system
32. Highest dry recovery along with curcumin is found in:
    a) Rashmi b) Suroma
    c) Ranga d) Sugandham
33. Bacterial gummosis is a common disease of:
    a) Citrus b) Bael
    c) Peach d) Apple
34. A chemical used as a preservative for coloured fruit is:
    a) Sodium benzoate b) KMS
    c) Sugar d) Salt
35. Freezing of fruits with cryogenic liquids is done at
    a) -200 b) -196
    c) 0 d) -4
36. Alcoholic fermentation is preferred for:
    a) Beverages b) Vinegar
    c) Pickles d) All of these
37. Two-dimensional diagrams represent the data by:
    a) Lines b) Simple bars
    c) Pie diagrams d) Cubes
38. Which fruit is having chromosome number 2n=56?
    a) Pineapple b) Mango
    c) Jackfruit d) Guava
39. Cyathium is a special type of inflorescence found in:
    a) *Euphorbia splendens* b) *Bougainvillea* spp.
    c) *Tradescantia* spp. d) *Massena* spp.

40. In India, horticulture sector covers only:
    a) 15% of total crop area  b) 20% of total crop area
    c) 8% of total crop area  d) 10% of total crop area

41. The equation of Hill reaction is:
    a) 6CO2+12 H2O – C6H12O6+6H2O + 6O2
    b) 2H2O + 2A – 2AH2 + O2
    c) C6H12O6 + 6O2 – 6 CO2 + 6H2O
    d) None of these

42. *Pyrostegia venusta* is an ornamental plant grouped as:
    a) Tree  b) Shrub
    c) Climber  d) Bulbous plant

43. Which one is a multi-bracted variety of bougainvillea?
    a) Roseville  b) Mahara
    c) Cherry  d) All of these

44. Litchi is a:
    a) Tropical evergreen fruit  b) Tropical deciduous fruit
    c) Subtropical deciduous fruit  d) Subtropical evergreen fruit

45. "Langra", a variety of mango is:
    a) A mid-season variety  b) An early variety
    c) A late variety  d) Ripens throughout the year

46. Foliar application of boron and zinc:
    a) Reduces fruit abscission  b) Increases fruit abscission
    c) Reduces fruit set  d) Does not affect abscission

47. The lowest level of inbreeding depression is observed in:
    a) Carrot  b) Onion
    c) Cucurbits  d) Cole crops

48. Bt. Brinjal, a genetically engineered variety of eggplant is resistant to:
    a) Little leaf  b) Bacterial wilt
    c) Phomopsis blight  d) Fruit and shoot borer

49. Reclamation of alkali soils is achieved by the application of:
   a) Lime b) Gypsum
   c) Magnesium chloride d) Calcium chloride
50. Which one of the following crops is affected by golden nematode?
   a) Cowpea b) Potato
   c) Cabbage d) Cauliflower
51. Which one of the following is a man-made vegetable?
   a) Jerusalem artichoke b) Rutabaga
   c) Garden peas d) Faha bean
52. Which is the largest orchid genus?
   a) Bulbophyllum b) Aerides
   c) Cypripedium d) Dendrobium
53. How much grass seed is required to establish 1 ha of lawn?
   a) 25 kg/ha b) 250g/ha
   c) 2.5 kg/ha d) All of the above
54. The horticultural crop production in India is:
   a) 185 MT b) 197 MT
   c) 207 MT d) 215 MT
55. The canes and the spurs left for flowering after pruning are called:
   a) Fruiting bodies b) Fruiting berries
   c) Fruiting clusters d) Modified flowers
56. What are the essential features of a modern English garden?
   a) Lawn b) Herbaceous border
   c) Rockery d) All of these
57. Which one of the following is an early variety of ber?
   a) Umram b) Thar Sevika
   c) Sansaur d) Banarasi Kadaka
58. Cytoplasmic male sterility is common in which of the following annuals?
   a) Ageratum d) Petunia
   c) Sunflower d) All of these

59. Calyx splitting in carnation is due to:
    a) The fluctuation of nigh temperature
    b) High level of potassium
    c) High level of nitrogen
    d) All of these
60. Which one of the following is the most common oxygenator plant in a water garden?
    a) Elodeal b) Crowfoot
    c) Victoria d) Typhus
61. Nugget is a hybrid cultivar of:
    a) Zinnia b) Marigold
    c) Antirrhinum d) Petunia
62. Bougainvillea climbs using which of the following:
    a) Tendrils b) Hooks
    c) Petiole d) All of these
63. Boron deficiency in pomegranate causes:
    a) Fruit drop b) Drying leaves
    c) Fruit cracking d) Onset of senescence
64. Bitterness in almond is mainly due to:
    a) Tartaric acid
    b) Citric acid
    c) *Cyanogenic glycosides* amygdalin
    d) Juglone
65. Mountain papaya is:
    a) *Carica parviflora* L. d) *Carica candamarcensis* hooker
    c) *Carica papaya* L. d) *Carissa macrocarpa* Jack
66. Wheat is native of:
    a) Central Asia (Turkey) b) Eastern
    c) Mexico d) U.S.A.
67. The ideal weight of date palm sucker for plantation is:
    a) 10-20 kg each b) 25-30 kg each
    c) 5-8 kg each d) 1-1.5 kg each

68. The ideal storage temperature for banana is:
    a) 11-20°C with RH less than 70%
    b) 11-20°C with RH more than 80%
    c) 5-8°C with RH less than 70%
    d) 11-20°C with RH more than 70%

69. Seedless watermelon was developed through:
    a) Mutation breeding  b) Transgenic approach
    c) Polyploidy breeding  d) Resistance breeding

70. The growth regulator commonly used as weedicide:
    a) NAA  b) ABC
    c) 2, 4-D  d) IBA

71. Date palm is native of:
    a) India  b) Mediterranean region
    c) Mexico  d) South Africa

72. The main objective of the Bahar treatment in fruit crop is to:
    a) Increase the vigour of the plant  b) Regulate the crop
    c) Deregulate the crop  d) Reduce the flowering

73. Seed plot technique has been developed in:
    a) Onion  b) Carrot
    c) Potato  d) Cole crops

74. Green leafy vegetables are a rich source of:
    a) Malic acid  b) Folic acid
    c) Citric acid  d) All of these

75. Wind pollination is common in:
    a) Watermelon  b) Cucumber
    c) Tomato  d) Spinach

76. Mode of pollination in spinach is:
    a) Cross-pollination  b) Often cross-pollination
    c) Self-pollination  d) It is cross incompatible

77. Which instrument is used t measure the Brix of jelly?
    a) Jelly thermometer  b) Brix meter
    c) Jel meter  d) Refractometer

78. Which of the following is a flower preservative?
    a) HgCl2 b) KNO3
    c) KMnO4 d) None of these
79. Which vegetable crop occupies the largest area in India?
    a) Potato b) Cole crops
    c) Tomato d) Onion
80. Brinjal originated in:
    a) India b) Bangladesh
    c) Europe d) Pakistan
81. Which floral part of saffron is used as a spice?
    a) Corolla b) Another
    c) Calyx d) Stigma
82. 'Cricket ball' is a variety of:
    a) Mango b) Sapota
    c) Guava d) Litchi
83. Vegetables are a rich source of:
    a) Protein b) Carbohydrate
    c) Vitamins d) Oil
84. IIVR is located at:
    a) Bikaner b) Varanasi
    c) Bhopal d) Cuttack
85. Rice variety IR-8 was introduced in India from:
    a) Mexico b) Canada
    c) USA d) Philippines
86. The ideal temperature of wheat tillering is:
    a) 20-30°C b) 16-20°C
    c) 10-15°C d) 23-25°C
87. Khaira disease in rice is caused by:
    a) Fungus b) Bacteria
    c) Virus d) None of these
88. The most serious pest of Bengal gram is:
    a) Cutworm b) Aphid
    c) Pod borer d) Smut

89. Which form of nitrogen is available in urea?
   a) Nitrate b) Ammonical
   c) Amide d) Nitrite
90. The percentage of oil contained in the Soybean seed is:
   a) 30% b) 50%
   c) 20% d) 40%
91. Hill soils are generally:
   a) Alkaline b) Acidic
   c) Neutral d) Saline
92. Triploids are generally:
   a) Fertile b) Sterile
   c) Male fertile d) Male sterile
93. Ribosomes are produced in:
   a) Mitochondria b) Endoplasmic reticulum
   c) Lysosomes d) Protoplast
94. TSS of jam should be:
   a) 70% b) 45%
   c) 35% d) 68%
95. As the fruit maturity progresses, the specific gravity:
   a) Increases b) Decreases
   c) Remain the same d) None of these
96. Which gas is emitted by ripening fruits?
   a) Methane b) Ethylene
   c) Ammonia d) Hexane
97. Which one of the citrus types is monoembryonic?
   a) Mandarin b) Pummelo
   c) Sour limed d) Citron
98. The highest production share among spice crops in India is of:
   a) Garlic b) Turmeric
   c) Chilli d) Ginger
99. Fruit crop suitable for high pH is:
   a) Apple b) Date palm
   c) Pomegranate d) Papaya

100. Which growth regulator is used for fruit thinning?
   a) IAA  b) NAA
   c) Cytokinins  d) GA3

101. The edible part of pomegranate is:
   a) Mesocarp  b) Aril
   c) Thalamus  d) Endocarp

102. Pineapple wilt is transmitted by:
   a) Whitefly  b) Aphids
   c) Mealybug  d) Lacewing bug

103. 'Arka Ananya' is a variety of:
   a) Brinjal  b) Tomato
   c) Bitter gourd  d) Onion

104. N-53 is a variety of:
   a) Spine gourd  b) Onion
   c) Ash gourd  d) Brinjal

105. The inflorescence type is collocation is:
   a) Panicle  b) Spadix
   c) Catkind  d) Umbel

106. Which of the following mechanisms exists in brinjal?
   a) Protoandry  b) Heterostyly
   c) Protogyny  d) Chasmogamy

107. Tip bending is a problem of:
   a) Rose  b) Gladiolus
   c) Orchid  d) Tuberose

108. The hermaphrodite variety of ridge gourd is:
   a) CO-1  b) Satputia
   c) Pusa Nutan  d) Hinga Sorai

109. Optimum temperature for tuber production in potato is:
   a) 10-15%  b) 15-20%
   c) 20-25%  d) 25-30%

110. Which type of tomatoes are called 'Self topping'?
   a) Intermediate  b) Indeterminate
   c) Semi-determinate  d) Determinate

111. Tenderometer is used to measure the maturity of:
   a) Cowpea b) Peapod
   c) French bean d) Dolichos bean
112. Which type of the following cucurbits exhibits vivipary?
   a) Cucumber b) Chayote
   c) Muskmelon d) Pumpkin
113. Whiptail in cauliflower can be corrected by:
   a) Addition of gypsum b) Soil liming
   c) Application of boron d) Late sowing
114. Vegetable with all edible plant parts is:
   a) Broad bean b) Winged bean
   c) Radish d) Lima bean
115. Which vegetable is the richest source of vitamin A?
   a) Drumstick leaves b) Carrot
   c) Pumpkin d) Pea
116. The optimum temperature for seed germination in pea is:
   a) 10°C b) 15°C
   c) 22°C d) 25°C
117. Oxalates are present in high amounts in:
   a) Root vegetables b) Leafy vegetables
   c) Pumpkin d) Alliums spp.
118. Softwood grafting is mainly adopted in:
   a) Aonla b) Bael
   c) Citrus d) Cashew nut
119. Which of the following species of Jasminum is not a climber?
   a) *J. grandiflorum* b) *J. dispermum*
   c) *J. humile* d) *J. officinale*
120. 'Thimma' is an important cultivar of:
   a) Rose b) Bougainvillea
   c) Tulip d) Tuberose
121. Seeds of which flower germinate only in dark:
   a) Marigold b) Nigella
   c) Zinnia d) Aster

122. Which of these is a day-neutral flower?
   a) Campanula b) Rudbeckia
   c) Chrysanthemum d) Gomphrena

123. Crossing-over takes place in:
   a) Laptotene b) Pachytene
   c) Diplotene d) Zygotene

124. The virus-free planting material can be produced by:
   a) Petiole b) Midrib
   c) Internode d) Meristem cultures

125. Which of the following is a ripening hormone?
   a) Auxin b) Ethylene
   c) Cycocel d) Cytokinin

126. Correlation coefficient lies between:
   a) 0 to 1.0 b) 0.1 to 1.0
   c) -1 to 1.0 d) 1.0 to 2.0

127. Which method is used in a small sample test?
   a) Chi-square test b) t-test
   c) F-test d) Z-test

128. Dormancy in potato is probed by:
   a) MH b) Alar
   c) Thiourea d) Cycocel

129. Zero energy cool chamber operates on the principle of:
   a) Charle's law b) Evaporative cooling
   c) Boyle's law d) Second law of thermodynamics

130. Calyx splitting is a disorder of:
   a) Dahlia b) Lily
   c) Carnation d) Chrysanthemum

131. A recently developed variety of Basmati rice is:
   a) Pusa 1450 b) Pusa 1121
   c) Pusa Sughandh d) Pusa Basmati

132. Karnal bunt is found in:
   a) Wheat b) Mango
   c) Pea d) Rice

133. Which is the cheapest preservation method of fruits and vegetables?

a) Drying b) Canning

c) Freezing d) Fermentation

134. Which of the following is useful for a diabetic patient?

a) Aonla b) Jamun

c) Avocado d) Grape

135. Which of the following is exposed to vapour heat treatment?

a) Apple b) Pomegranate

c) Mango d) Pear

136. For distant markets, the harvesting of melons is done at:

a) Mature stage b) Half-slip stage

c) Full slip stage d) Green mature stage

137. Which of the following is a n Agricultural marketing organization?

a) AVRDC b) AADF

c) FCI d) NAFED

138. Commercial mango ripening is induced by using:

a) NAA b) Calcium carbide

c) Acetylene d) Ethrel

139. Which of the following is a powder mildew resistant garden pea variety?

a) Bonneville b) Lincoln

c) Palam Priya d) Azad P1

140. Disbudding is a common practice in:

a) Jasmine b) Gerbera

c) Chrysanthemum d) Rose

141. *Delonix regia* is commonly known as:

a) Bottlebrush b) Gulmohar

c) Palas d) Jacaranda

142. Which of the following is the major loose-flower producing state of India?

a) Karnataka b) Andhra Pradesh

c) Tamil Nadu d) Maharashtra

143. The inflorescence of cruciferous family is known as:

a) Catkin b) Raceme

c) Umbel d) None of these

144. An example of cleistogamous vegetable is:

| | | | |
|---|---|---|---|
| a) | Golden pea | b) | Lettuce |
| c) | Capsicum | d) | Tomato |

145. An example of autotetraploid vegetable is:

| | | | |
|---|---|---|---|
| a) | Leek | b) | Potato |
| c) | Okra | d) | Artichoke |

146. Which one of the following is a yellow vein mosaic resistant variety of okra?

| | | | |
|---|---|---|---|
| a) | Pusa Sawani | b) | Harbhajan |
| c) | Arka Anamika | d) | Pusa Makhmali |

148. The plant known as sweetener is:

| | | | |
|---|---|---|---|
| a) | Rose | b) | Jasmine |
| c) | Aloe | d) | Stevia |

149. Gynoecious variety of papaya is:

| | | | |
|---|---|---|---|
| a) | Pusa Giant | b) | Pusa Dwarf |
| c) | Sunrise Solo | d) | Pusa Nanha |

150. Tomato leaf curl virus is transmitted by:

| | | | |
|---|---|---|---|
| a) | Aphids | b) | White fly |
| c) | Hopper | d) | Thrips |

151. Matching the following:

| | | | |
|---|---|---|---|
| i) | Antirrhinum | a) | Amaryllis |
| ii) | Spider lily | b) | Paper flower |
| iii) | Sword lily | c) | Dog flower |
| iv) | Acroclinium | d) | Ice plant |
| v) | Mesembryanthemum | e) | Gladiolus |

152. Matching the following:

| | | | |
|---|---|---|---|
| i) | Pusa Sambandh | a) | Molybdenum |
| ii) | Pusa Yamdagini | b) | Cauliflower |
| iii) | Whiptail | c) | Cabbage |
| iv) | Browning | d) | Culling |
| v) | Sweet potato | e) | Carrot |

153. Match The Following

| | | | |
|---|---|---|---|
| i) | Mango | a) | Polyembroyon |
| ii) | Morcottage | b) | Temperature Litchi |
| iii) | Olur | c) | Litchi |
| iv) | Kiwi fruit | d) | Exotic |
| v) | Guava | e) | India |

154. Match The Following

| | | | |
|---|---|---|---|
| i) | Banana | a) | Tas-e-Ganesh |
| ii) | Grape | b) | Shade-loving |
| iii) | Dieffenbachia | c) | Curcumin |
| iv) | Turmeric | d) | Climber |
| v) | Dens scandens | e) | Pseudostem |

155. Match The Following

| | | | |
|---|---|---|---|
| i) | Dieback in litchi | a) | Boron |
| ii) | Improper banana finger filling | b) | Copper |
| iii) | Calyx end rot of grape | c) | Zinc |
| iv) | Water core of apple | d) | Calcium |
| v) | Leaf bronzing of litchi | e) | Potash |

156. Match The Following

| | | | |
|---|---|---|---|
| i) | St. George | a) | Walnut |
| ii) | Khimi | b) | Apricot |
| iii) | Rancidity | c) | Acid lime |
| iv) | Brown rot | d) | Sapota |
| v) | Cross protection | e) | Grape |

157. Match The Following

| | | | |
|---|---|---|---|
| i) | English garden | a) | Wells |
| ii) | Japanese garden | b) | Lawn |
| iii) | Mughal garden | c) | Baradari |
| iv) | Garden adornment | d) | Seat |
| v) | Garden style | e) | Informal |

158. Match The Following

| | | | |
|---|---|---|---|
| i) | Marigold | a) | Bulb |
| ii) | Tuberose | b) | Cutting |
| iii) | Carnation | c) | Rhizome |
| iv) | Canna | d) | Seed |
| v) | Gladiolus | e) | Corm |

159. Match The Following

| | | | |
|---|---|---|---|
| i) | Ridge gourd | a) | 22 |
| ii) | French bean | b) | 26 |
| iii) | Okra | c) | 12 |
| iv) | Pea | d) | 130 |
| v) | Spinach | e) | 14 |

160. Match The Following

| | | | |
|---|---|---|---|
| i) | Linalool | a) | Papaya |
| ii) | Curcumin | b) | Coriander |
| iii) | Diosgenin | c) | Fenugreek |
| iv) | Papin | d) | Fig |
| v) | Ficin | e) | Turmeric |

## Answer Key

| | | | | | | | | | | | | | |
|---|---|---|---|---|---|---|---|---|---|---|---|---|---|
| 1. | (a) | 2. | (d) | 3. | (d) | 4. | (c) | 5. | (a) | 6. | (d) | 7. | (b) |
| 8. | (a) | 9. | (a) | 10. | (c) | 11. | (a) | 12. | (a) | 13. | (a) | 14. | (d) |
| 15. | (a) | 16. | (c) | 17. | (d) | 18. | (c) | 19. | (a) | 20. | (b) | 21. | (b) |
| 22. | (c) | 23. | (d) | 24. | (a) | 25. | (d) | 26. | (c) | 27. | (c) | 28. | (b) |
| 29. | (a) | 30. | (a) | 31. | (c) | 32. | (b) | 33. | (c) | 34. | (a) | 35. | (b) |
| 36. | (b) | 37. | (c) | 38. | (c) | 39. | (a) | 40. | (d) | 41. | (a) | 42. | (c) |
| 43. | (d) | 44. | (d) | 45. | (a) | 46. | (a) | 47. | (c) | 48. | (d) | 49. | (a) |
| 50. | (b) | 51. | (b) | 52. | (b) | 53. | (a) | 54. | (a) | 55. | (d) | 56. | (a) |
| 57. | (d) | 58. | (b) | 59. | (d) | 60. | (a) | 61. | (a) | 62. | (b) | 63. | (b) |
| 64. | (c) | 65. | (c) | 66. | (b) | 67. | (a) | 68. | (a) | 69. | (b) | 70. | (c) |
| 71. | (c) | 72. | (b) | 73. | (b) | 74. | (c) | 75. | (b) | 76. | (d) | 77. | (a) |
| 78. | (d) | 79. | (a) | 80. | (a) | 81. | (a) | 82. | (d) | 83. | (b) | 84. | (c) |
| 85. | (b) | 86. | (d) | 87. | (b) | 88. | (d) | 89. | (c) | 90. | (c) | 91. | (c) |
| 92. | (b) | 93. | (b) | 94. | (b) | 95. | (d) | 96. | (a) | 97. | (b) | 98. | (b) |

| | | | | | | | | | | | | | |
|---|---|---|---|---|---|---|---|---|---|---|---|---|---|
| 99. | (c) | 100. | (b) | 101. | (b) | 102. | (b) | 103. | (c) | 104. | (b) | 105. | (b) |
| 106. | (b) | 107. | (b) | 108. | (b) | 109. | (b) | 110. | (b) | 111. | (d) | 112. | (b) |
| 113. | (b) | 114. | (b) | 115. | (b) | 116. | (b) | 117. | (c) | 118. | (b) | 119. | (d) |
| 120. | (c) | 121. | (b) | 122. | (b) | 123. | (d) | 124. | (b) | 125. | (d) | 126. | (b) |
| 127. | (c) | 128. | (b) | 129. | (c) | 130. | (b) | 131. | (c) | 132. | (b) | 133. | (a) |
| 134. | (a) | 135. | (b) | 136. | (c) | 137. | (b) | 138. | (d) | 139. | (d) | 140. | (c) |
| 141. | (c) | 142. | (b) | 143. | (c) | 144. | (b) | 145. | (b) | 146. | (b) | 147. | (c) |
| 148. | (d) | 149. | (c) | 150. | (b) | | | | | | | | |

| Q. | 151. | 152. | 153. | 154. | 155. | 156. | 157. | 158. | 159. | 160. |
|---|---|---|---|---|---|---|---|---|---|---|
| i) | (c) | (c) | (e) | (e) | (b) | (e) | (b) | (d) | (b) | (b) |
| ii) | (a) | (e) | (c) | (a) | (e) | (d) | (a) | (a) | (a) | (e) |
| iii) | (e) | (a) | (a) | (b) | (d) | (a) | (c) | (b) | (d) | (c) |
| iv) | (b) | (b) | (b) | (c) | (a) | (b) | (d) | (c) | (e) | (a) |
| v) | (d) | (d) | (d) | (d) | (c) | (c) | (e) | (e) | (c) | (d) |

# 2

# ICAR – JRF Horticulture Exam – 2011

1. Horticulture sector covers % of total cropped area.
   a) 10% b) 15%
   c) 20% d) 25%
2. World Trade Organization is situated at:
   a) Geneva b) Rome
   c) Paris d) Brazil
3. AVRDC is situated at:
   a) Phillipines b) Taiwan
   c) Japan d) India
4. Which of the following International organizations is located in India.
   a) IRRI b) CIMMYT
   c) ICRISAT d) AVRDC
5. The simplest measure of dispersion in a data set is:
   a) Range b) Standard deviation
   c) Variance d) Mean deviation
6. Metaxenia is normally found in which of the following fruit crops:
   a) Aonla b) Ber
   c) Date palm d) Fig
7. The fruit type of pomegranate is:
   a) Hesperidium b) Capsule
   c) Balausta d) Syconus
8. Inter-stock is used to overcome:
   a) Sexual incompatibility b) Graft incompatibility
   c) Stionic effect d) None of these

9. Development of embryo from unfertilized egg cell is known as:
   a) Apomixis b) Parthenocarpy
   c) Parthenogenesis d) Apogamy
10. Which of the following is a micronutrient loving plant?
   a) Banana b) Papaya
   c) Citrus d) Grape
11. 'Indian Horticulture' is a semi-technical magazine published from ICAR as:
   a) Monthly b) Bimonthly
   c) Quarterly d) Half-yearly
12. "Advances in Arid Horticulture" has been written by:
   a) K.L. Chadha b) O.P. Pareek
   c) O.P. Awasthi d) M.S. Randhawa
13. Gummosis in citrus is caused by:
   a) Virus b) Bacteria
   c) Fungus d) Nutrient deficiency
14. The non-preference insect mechanism is known as:
   a) Antiphenromonasis b) Antibiotic
   c) Antichlorosis d) Antixenosis
15. Which of the following fruit has the highest amount of riboflavin (vitamin B2)?
   a) Ber b) Litchi
   c) Apricot d) Bael
16. The ideal planting time for grape in northern India is:
   a) July b) October
   c) January d) March
17. The PGR used as an alternative to the chilling requirement in many temperate fruits is:
   a) GA3 b) NAA
   c) Kinetin d) ABA
18. Cashew nut originated in:
   a) Mexico b) Peru
   c) Brazil d) Paraguay

19. IISR is located at:
    a) Bengaluru  b) Calicut
    c) Varanasi  d) Ajmer
20. The range of correlation coefficient is:
    a) 0 to 1  b) -1 to +1
    c) -1 to 0  d) 0 to -1
21. Which of the following is used as an intercrop in ber orchard?
    a) Mustard  b) Aloe
    c) Groundnut  d) Wheat
22. "Chicken Tongue" is related to:
    a) Grape  b) Litchi
    c) Papaya  d) Cherry
23. The fruit crop known as "King of Temperate Fruits" is:
    a) Apricot  b) Cherry
    c) Apple  d) Pear
24. Degreening is carried out in most fruit crops with the exception of:
    a) Guava  b) Citrus
    c) Banana  d) Mango
25. Simultaneous growing of two or more crops by mixing their seeds is known as:
    a) Mixed farming  b) Intercropping
    c) Mixed cropping  d) Alley cropping
26. In double-helical DNA structure, thymine always pairs with:
    a) Adenine  b) Guanine
    c) Cytosine  d) Uracil
27. Cultivated strawberry (*Fragaria × ananasa*) is a natural hybrid of:
    a) Fragaria chiloensis × Fragaria vesca
    b) Fragaria chiloensis × Fragaria virginiana
    c) Fragaria virginiana × Fragaria vesca
    d) Fragaria virginiana × Fragaria chiloensis
28. The flower colour of 'Kasuri Methi' is:
    a) White  b) Purple
    c) Yellow  d) Green

29. The fruit crop highly susceptible to waterlogging is:
    a) Grape b) Banana
    c) Pineapple d) Papaya
30. Ca deficiency symptoms first appear on:
    a) Younger leaves b) Older leaves
    c) Terminal bud d) None of these
31. Examples of non-climacteric fruit crops are:
    a) Banana and mango b) Guava and orange
    c) Mandarin and grape d) All of these
32. Turning brown of apple flesh on exposure to air is caused due to the action of:
    a) Lipoxygenase b) Catalase
    c) Ascorbic acid oxidase d) Polyphenol oxidase
33. Mango hybrid 'Ratna' is a product of a cross between:
    a) Neelum × Mallika b) Mallika × Neelum
    c) Alphonso × Neelum d) Neelum × Alphonso
34. Somatic chromosome number in *Musa sapientum* is
    a) 44 b) 22
    c) 33 d) 66
35. Dioecious form of sex is found in:
    a) Date palm b) Asparagus
    c) Papaya d) All of these
36. Which of the following is used to control fruit drop in citrus?
    a) 2,4 – D b) ABA
    c) IAA d) GA3
37. Which of the following is a polygamous fruit crop?
    a) Guava b) Mango
    c) Papaya d) Mandarins
38. Nectarines are:
    a) Smooth skinned peaches b) Dried almonds
    c) Smooth skinned plums d) Smooth skinned apricots

39. Which of the following fruits has the highest productivity in India?
    a) Banana b) Papaya
    c) Mango d) Pineapple
40. Which of the following fruits is the best source of jelly?
    a) Guava b) Mango
    c) Banana d) Orange
41. Guava is a rich source of:
    a) Protein b) Fibre
    c) Fat d) Sugar
42. Which of the following is known as Gulab jamun or Rose Apple?
    a) Syzygium cumini b) Syzygium fruticosum
    c) Syzygium jambos d) Psidium guajava
43. Which of the following is not a common method to train grapes?
    a) Central leader b) Telephone
    c) Kniffin d) Head
44. For resin making, grapes are dipped in:
    a) Copper sulphate b) Sodium hydroxide
    c) Sulphuric acid d) CCC
45. How much percentage of male plants is required in papaya orchard?
    a) 5 b) 15
    c) 10 d) 20
46. Loquat (*Eriobotrya japonica*) is a member of .........family:
    a) Liliaceae b) Rosaceae
    c) Apocynaceae d) Vitaceae
47. What percentage of polliniser cultivar should be planted in an apple orchard?
    a) 15 b) 25
    c) 33 d) 50
48. Which of the following is a cultivar of apricot?
    a) Poovan b) Mallika
    c) Chandler d) New Castle
49. For stratification, the seeds are kept in the sand at a temperature of
    a) 1-3°C b) 4-7°C
    c) 8-11°C d) 12-22°C

50. Apple scab is caused by:
    a) Venturia b) Alternaria
    c) Fusarium d) Botrytis
51. Fruit cracking in pomegranate can be controlled by spraying:
    a) Iron b) Boron
    c) Calcium d) Phorate
52. Cryopreservation is associated with:
    a) Liquid nitrogen b) Liquid oxygen
    c) Liquid potassium d) Liquid carbon dioxide
53. Percentage of oil in mint is:
    a) 0.1-0.5 b) 0.5-1.0
    c) 1-1.5 d) 2-3
54. Winter banana is a variety of:
    a) Mango b) Apple
    c) Banana d) Pear
55. Propping is done in:
    a) Apple b) Sapota
    c) Banana d) Bael
56. Which of the following fruit is exotic?
    a) Pineapple b) Bael
    c) Mango d) Phalsa
57. How many more trees are planted in the equilateral triangle system of planting,?
    a) 30% b) 25%
    c) 15% d) 5%
58. Root and shoot pruning is done in:
    a) Jamun b) Papaya
    c) Gladiolus d) Roses
59. Nucellar seedlings can be used to raise true to type plants in:
    a) Kagzi lime b) Cashewnut
    c) Litchi d) Coconut

60. Scarification in guava is done by keeping seeds in hot water at a temperature range of:
    a) 60-70°C b) 77-100°C
    c) 70-80°C d) 50-60°C
61. Stooling is being practiced in:
    a) Guava b) Loquat
    c) Banana d) Aonla
62. Epicotyl grafting was commercialized for rapid multiplication of mango in:
    a) Andhra Pradesh b) Maharashtra
    c) Gujarat d) Madhya Pradesh
63. Inter stock is used to overcome:
    a) Asexual incompatibility b) Stionic effect
    c) Inter sterility d) Sexual incompatibility
64. tissue culture is commercially exploited in which of the following fruit crops?
    a) Apple b) Mango
    c) Aonla d) Banana
65. Apple may be stored at which optimal storage condition for 4-8 months?
    a) 2 to 2.5°C temp and 80-85% RH
    b) 1 to 2°C temp and 80-85% RH
    c) -1 to 0°C temp and 85-90% RH
    d) -0.8 to 1°C temp and 95-100% RH
66. Caprification is generally done in which fruit crop?
    a) Wild fig b) Symrna fig
    c) Edible fig d) Common fig
67. Best quality guavas are produced in:
    a) Hasth bahar b) Mrig bahar
    c) Ambe bahar d) Mixed bahar
68. Most suitable fruit crop for pickle making is:
    a) Spondias pinnata b) Malpighia hispida
    c) *Annona squamosa* d) *Limonia acidissima*
69. Kinnow is the first-generation hybrid of:
    a) *Citrus sinensis* × *C. paradisi* b) *Citrus reticulata* × *C. limettioides*
    c) *Citrus nobilis* × *C. deliciousa* d) *Citrus sinensis* × *C. aurantifolia*

70. Gynodioecious variety of papaya is:
    a) Pusa Nanha b) Pusa Giant
    c) Sunrise Solo d) Pant-1
71. Sigatoka is an important disease of:
    a) Strawberry b) Guava
    c) Banana d) Cashew nut
72. Bacterial gummosis is a common disease of:
    a) Citrus b) Bael
    c) Peach d) Apple
73. A chemical used as a preservative for coloured fruit is:
    a) Sodium benzoate b) KMS
    c) Sugar d) Salt
74. chromosome 2n=56 is found in which of the following fruits ?
    a) Pear b) Mango
    c) Jackfruit d) Guava
75. "Goma Aishwarya" is a variety of:
    a) Rose b) Aonla
    c) Tomato d) Kalmegh
76. The edible part of citrus fruit is:
    a) Fleshy receptacle b) Endocarp
    c) Mesocarp d) Juicy placenta
77. Pheromones trap attracts:
    a) Caterpillar b) Female bugs
    c) Male d) Female moths
78. "Pusa Srijan" is a rootstock of:
    a) Citrus b) Guava
    c) Apple d) Mango
79. Gladiolus belongs to the family:
    a) Iridaceae b) Rosaceae
    c) Compositae d) Solanaceae
80. An optimum seed rate of onion is:
    a) 4-5 kg/ha b) 15-20 kg/ha
    c) 8-10 kg/ha d) 20-25 kg/ha

81. Which of the following causes pungency in radish?
    a) Diallyl disulphide  b) Isothiocyanate
    c) Allyl propyl disulphide  d) Erucic acid
82. Leaf curl virus of chilli is transmitted by:
    a) Jassids  b) Aphid
    c) Whitefly  d) Housefly
83. Buttoning is a physiological disorder of:
    a) Garlic  b) Cabbage
    c) Carrot  d) Cauliflower
84. The edible part of Brussels sprouts:
    a) Root  b) Small head or bud
    c) Seed  d) Leaves
85. About 95% to 99.5% portion of plant tissue is made of:
    a) C, H and O  b) N, P and K
    c) Ca, Mg and C  d) CU, Zn and Fe
86. The pigment predominantly responsible for the red colour of tomatoes is:
    a) Carotene  b) Xanthophyll
    c) Anthocyanin  d) Lycopene
87. The golden nematode is a common pest of:
    a) Tomato  b) Okra
    c) Brinjal  d) Lima bean
88. The often cross-pollinated vegetable crop is:
    a) Peas  b) Tomato
    c) Brinjal  d) Lima bean
89. "Blackleg" is an important disease of:
    a) Cauliflower  b) Cabbage
    c) Radish  d) Carrot
90. 'Arka Alok', a tomato variety, is resistant to:
    a) Wilt  b) Nematode
    c) Fruit borer  d) High-temperature stress
91. Chimera can be maintained by:
    a) Somatic embryogenesis  b) Apomixis
    c) Adventitious shoot formation  d) Cutting and grafting

92. Chrysanthemum is a:
    a) Short day plant  b) Long day plant
    c) Day-neutral plant  d) None of these
93. Propagation of Lilium is done through:
    a) Tuber  b) Corm
    c) Rhizome  d) Bulb
94. Which of the following suffers from calyx splitting?
    a) Carnation  b) Marigold
    c) Rose  d) Chrysanthemum
95. The preservative used in tomato ketchup is:
    a) Potassium metabisulphite  b) Propionic acid
    c) Niacin  d) Sodium benzoate
96. 2, 4-D is a/an:
    a) Auxin  b) Fungicide
    c) Growth hormone  d) Insecticide
97. Which of the following is a late-maturing variety of cauliflower?
    a) Pusa Himjyoti  b) Pusa Deepali
    c) Pusa Synthetic  d) Pusa Katki
98. Date palm is propagated using:
    a) Rhizome  b) Suckers
    c) Seeds  d) Runners
99. The bulbils in garlic develop naturally from modified:
    a) Fruits  b) Shoots
    c) Flowers  d) Roots
100. Bougainvillaea climbs using:
    a) Tendrils  b) Hooks
    c) Petiole  d) All of these
101. Which floral part of saffron is used as a spice?
    a) Corolla  b) Anther
    c) Calyx  d) Stigma
102. Which of the following is known as vinegar bacteria?
    a) Bacillus spp.  b) Clostridium
    c) Acetobacter  d) Lactobacillus

103. Seed plot technique has been developed for:
a) Onion b) Carrot
c) Potato d) Cole crops

104. "French Breakfast" comprises a variety of:
a) Radish b) Carrot
c) Turnip d) Broccoli

105. The lowest level of inbreeding depression is observed in:
a) Carrot b) Onion
c) Cucurbits d) Cole crops

106. Which family does sweet potato belong to ?:
a) Rosaceae b) Euphorbiaceae
c) Convolvulaceae d) Cruciferae

107. The edible part in broccoli is known as:
a) Curd b) Flower bud
c) Flower stalk d) Head

108. Sulphur based fungicide can be used on all vegetables except:
a) Cucurbits b) Crucifers
c) Root vegetables d) Okra

109. The term bonsai was popularized by:
a) Korean b) Chinese
c) Japanese d) American

110. Which of the following crop is not grown for perfumery?
a) Rose b) Lilium
c) Geranium d) Jasmine

111. Most of the companies that exportdry flowers are situated at:
a) Tuticorin b) Goa
c) Bangalore d) Shimla

112. Which of the following is a suitable desiccant for flower drying?
a) Silica gel b) Sand
c) Boric acid d) All of these

113. Nishkant; a thornless rose rootstock was developed at:
a) IARI b) NBRI
c) IIHR d) BARC

114. Which of the following is a hybrid of African marigold?
   a) Pusa Basanti b) Pusa Narangi
   c) Pusa Shankar - 1 d) Pusa Shweta
115. Grey mould in chrysanthemum is caused by:
   a) Fusarium b) Botrytis
   c) Alternaria d) Septoria
116. Gulkand is prepared by mixing rose petals and sugar in the ratio of:
   a) 1: 1 b) 1: 2
   c) 1 : 3 d) 2 : 1
117. Perpetual carnation is commercially propagated by:
   a) Terminal stem cutting b) Seeds
   c) Suckers d) All of these
118. Genetic male sterility is very common in:
   a) China aster b) Marigold
   c) Gaillardia d) Dahlia
119. Which of the following cole crops has the highest outcrossing?
   a) Cauliflower b) Kale
   c) Cabbage d) Broccoli
120. Browning in cauliflower is due to the deficiency of:
   a) Cu b) Ca
   c) Bo d) Mo
121. Heliconia is commercially propagated through:
   a) Bulb b) Rhizome
   c) Corm d) Tuber
122. Oxalates are predominant in:
   a) Leafy vegetables b) Root vegetables
   c) Bulb crops d) Fruit vegetables
123. Greening in potato is caused due to:
   a) High temperature b) High light intensity
   c) Low light intensity d) Low temperature
124. TSS of jam should not be more than:
   a) 50°B b) 60°B
   c) 70°B d) 80°B

125. In India, date is harvested at ..... stage:

a) Dang b) Pind

c) Doka d) Immature

126. Fruit of rose is known as:

a) Bulbs b) Hips

c) Caryopsis d) Berry

127. The quickest method of laying a lawn is:

a) Seedling b) Turfing

c) Dibbing d) Dung plastering

128. Malling Merton series of rootstock is related to:

a) Almond b) Apple

c) Strawberry d) Plum

129. "Thimma" is an important variety of:

a) Rose b) Bougainvillaea

c) Tuberose d) Tulip

130. The centre of origin of cocoa is:

a) Brazil b) South America

c) Asia d) Kenya

131. Randomization is done to remove:

a) Biasness b) Degree of freedom

c) LSD d) Probability

132. Which medicinal plant is used in controlling blood pressure?

a) *Withania somnifera* b) *Catharanthus roseus*

c) *Rauwolfia serpentina* d) *Aloe vera*

133. Among the following flowers, which one is most resistant to nematode infestation: vis:

a) Rose b) Jasmine

c) Gladiolus d) Marigold

134. Which one of the following is a major problem in coconut cultivation?

a) Lack of good varieties b) Eryophid mite

c) Stem bleeding d) Red palm weevil

135. Rashtrapati Bhawan Garden is an example of which type of gardens?

a) Mughal b) Japanese

c) English d) Italian

136. The cultivated turmeric is:

a) Diploid b) Tetraploid

c) Triploid d) Hexaploid

137. Photosynthesis occurs in:

a) Chloroplast b) Mitochondria

c) Golgi bodies d) Palms

138. Tissue culture is common in:

a) Rose b) Orchids

c) Gladiolus d) Palms

139. The pungency of garlic is due to:

a) Allyl propyl disulphide b) Isothiocyanate

c) Diallyl disulphide d) None of these

140. Curing is done in which one of the following vegetable crops?

a) Garlic b) Chilli

c) Onion d) Potato

141. Blossom end rot in tomato is caused due to:

a) Pathological disorder b) Viral disorder

c) Physiological disorder d) Genetical disorder

142. The hollow heart may appear in:

a) Small-sized tubers b) Oversized tubers

c) Medium-sized tubers d) None of these

143. "Chalta tree" is botanically known as:

a) *Derris scandens* b) *Dillenia indica*

c) *Delonix pulcherima* d) *Diospyros kaki*

144. Which flower is most suited for wreath arrangements?

a) White fragrant roses b) White gladiolus

c) Coloured roses d) Coloured gladiolus

145. Seeds contain important steroid diosgenin:

a) Fenugreek b) Fennel

c) Nutmeg d) Tamarind

146. Among the following, which medicinal plant is used for culinary purposes?

a) Calotropic  b) Adhatoda

c) Aloe  d) Safed Musli

147. Mushroom is a kind of edible............:

a) Bacterium  b) Algae

c) Fungus  d) Yeast

148. Rambutan belongs to ............family:

a) Rutaceae  b) Sapindaceae

c) Guttiferreae  d) Anacardiaceae

149. In which of the following crops is curing done for ripening?

a) Guava  b) Apple

c) Orange  d) Banana

150. Bt. brinjal has been developed for:

a) Root borer resistance  b) Enrichment of Vitamin A

c) Fruit and shoot borer resistance  d) Higher yield

151. Match the following:

| | | | |
|---|---|---|---|
| i) | Drought | a) | Measurement of stomatal aperture |
| ii) | Psychrometer | b) | Somatic hybridization |
| iii) | Porometer | c) | Measurement of tissue water |
| iv) | Protoplast fusion | d) | Abiotic stress |
| v) | Chilling requirement | e) | Temperature fruits |

152. Match the following:

| | | | |
|---|---|---|---|
| i) | Self-pollination | a) | Heterozygosity |
| ii) | Cross-pollination | b) | Homozygosity |
| iii) | Apomixis | c) | Sexual reproduction |
| iv) | Amphimixis | d) | Lack of sexual fusion |
| v) | Self-incompatibility | e) | Almond |

153. Match the following gardens (left) with their correct locations (right):

| | | | |
|---|---|---|---|
| i) | Rose Garden | a) | Darjeeling |
| ii) | Dilkusha Garden | b) | Japan |
| iii) | Roshanara garden | c) | New Delhi |
| iv) | Osaka Garden | d) | Lahore |
| v) | Lyod botanical | e) | Chandigarh |

154. Match the following:

| | | | |
|---|---|---|---|
| i) | Alkali soil | a) | Guava canker |
| ii) | Copper deficiency | b) | Vector |
| iii) | Fungus | c) | Guava wilt |
| iv) | Zinc deficiency | d) | Little leaf |
| v) | White fly | e) | Gummosis |

155. Match the following crops (left) with their respective families (right):

| | | | |
|---|---|---|---|
| i) | Bathua | a) | Chenopodiaceae |
| ii) | Carrot | b) | Umbelliferae |
| iii) | Pea | c) | Fabaceae |
| iv) | Okra | d) | Malvaceae |
| v) | Sweet potato | e) | Convolvulaceae |

156. Match the following:

| | | | |
|---|---|---|---|
| i) | Plum | a) | Stone fruit |
| ii) | July Elberta | b) | Peach |
| iii) | Guava canker | c) | Fungus |
| iv) | Little leaf in brinjal | d) | MLO |
| v) | Gummosis in peach | e) | Bacterium |

157. Match the following cell parts (left) with their corresponding functions (right):

| | | | |
|---|---|---|---|
| i) | Chloroplast | a) | Protein synthesis |
| ii) | Mitochondria | b) | Photosynthesis |
| iii) | Ribosomes | c) | Packaging of food material |
| iv) | Golgi bodies | d) | Cellular respiration |
| v) | Lysosomes | e) | Digestive vacuoles |

158. Match the following:

| | | | |
|---|---|---|---|
| i) | Colt | a) | Cherry |
| ii) | Flying Dragon | b) | Ultra dwarf rootstock of citrus |
| iii) | Severinia bouxifolia | c) | Salt tolerant rootstock of citrus |
| iv) | Quince C | d) | Pear |
| v) | Manilkara hexandra | e) | Sapota |

159. Match the following institutions (left) with their corresponding locations (right):

| | | | |
|---|---|---|---|
| i) | CAZRI | a) | New Delhi |
| ii) | CPRI | b) | Shimla |
| iii) | IGFRI | c) | Jhansi |
| iv) | NRC for Cashew | d) | Puttur |
| v) | IASRI | e) | Jodhpur |

160. Match the following:

| | | | |
|---|---|---|---|
| i) | 2, 4-D | a) | Weedicide |
| ii) | Auxin | b) | Vitamin |
| iii) | Malathion | c) | Insecticide |
| iv) | Thirum | d) | Fungicide |
| v) | Riboflavin | e) | Plant growth regulator |

## Answers Key

| | | | | | | | | | | | | | |
|---|---|---|---|---|---|---|---|---|---|---|---|---|---|
| 1. | (a) | 2. | (a) | 3. | (b) | 4. | (c) | 5. | (a) | 6. | (c) | 7. | (c) |
| 8. | (b) | 9. | (c) | 10. | (c) | 11. | (b) | 12. | (c) | 13. | (c) | 14. | (d) |
| 15. | (d) | 16. | (c) | 17. | (a) | 18. | (c) | 19. | (b) | 20. | (b) | 21. | (c) |
| 22. | (b) | 23. | (c) | 24. | (a) | 25. | (c) | 26. | (a) | 27. | (b) | 28. | (c) |
| 29. | (d) | 30. | (c) | 31. | (c) | 32. | (d) | 33. | (d) | 34. | (b) | 35. | (d) |
| 36. | (a) | 37. | (c) | 38. | (a) | 39. | (b) | 40. | (a) | 41. | (b) | 42. | (c) |
| 43. | (a) | 44. | (b) | 45. | (c) | 46. | (b) | 47. | (c) | 48. | (d) | 49. | (b) |
| 50. | (a) | 51. | (b) | 52. | (a) | 53. | (a) | 54. | (b) | 55. | (c) | 56. | (a) |
| 57. | (c) | 58. | (d) | 59. | (a) | 60. | (b) | 61. | (a) | 62. | (b) | 63. | (a) |
| 64. | (d) | 65. | (c) | 66. | (b) | 67. | (b) | 68. | (a) | 69. | (c) | 70. | (c) |
| 71. | (c) | 72. | (c) | 73. | (a) | 74. | (c) | 75. | (b) | 76. | (d) | 77. | (c) |
| 78. | (b) | 79. | (a) | 80. | (c) | 81. | (b) | 82. | (c) | 83. | (d) | 84. | (b) |
| 85. | (a) | 86. | (d) | 87. | (d) | 88. | (d) | 89. | (b) | 90. | (a) | 91. | (d) |
| 92. | (a) | 93. | (d) | 94. | (a) | 95. | (d) | 96. | (a) | 97. | (a) | 98. | (b) |
| 99. | (c) | 100. | (b) | 101. | (d) | 102. | (c) | 103. | (c) | 104. | (a) | 105. | (c) |
| 106. | (c) | 107. | (b) | 108. | (a) | 109. | (c) | 110. | (b) | 111. | (a) | 112. | (c) |
| 113. | (c) | 114. | (c) | 115. | (b) | 116. | (a) | 117. | (a) | 118. | (b) | 119. | (d) |
| 120. | (c) | 121. | (b) | 122. | (a) | 123. | (b) | 124. | (c) | 125. | (c) | 126. | (b) |
| 127. | (b) | 128. | (b) | 129. | (b) | 130. | (b) | 131. | (a) | 132. | (b) | 133. | (d) |
| 134. | (b) | 135. | (a) | 136. | (c) | 137. | (a) | 138. | (b) | 139. | (c) | 140. | (c) |
| 141. | (c) | 142. | (b) | 143. | (b) | 144. | (a) | 145. | (a) | 146. | (c) | 147. | (c) |
| 148. | (b) | 149. | (d) | 150. | (c) | | | | | | | | |

| Q. | 151. | 152. | 153. | 154. | 155. | 156. | 157. | 158. | 159. | 160. |
|---|---|---|---|---|---|---|---|---|---|---|
| i) | (d) | (b) | (e) | (c) | (a) | (a) | b) | (a) | (e) | (a) |
| ii) | (c) | (a) | (d) | (e) | (b) | (b) | d) | (b) | (b) | (e) |
| iii) | (a) | (d) | (c) | (a) | (c) | (c) | a) | (c) | (c) | (c) |
| iv) | (b) | (c) | (b) | (d) | (d) | (d) | c) | (d) | (d) | (d) |
| v) | (e) | (e) | (a) | (b) | (e) | (e) | e) | (e) | (a) | (b) |

# 3

# ICAR – JRF Horticulture Exam – 2012

1. Horticulture sector covers % of total cropped area.

   a) 10% b) 15%

   c) 20% d) 25%

2. Among all floricultural products, the contribution of dry flowers in total exports is:

   a) 40 b) 50

   c) 60 d) 70

3. Protein synthesis takes place in:

   a) Nucleus b) Nucleolus

   c) Ribosome d) Mitochondria

4. DNA synthesis takes place in:

   a) S – phase b) $G_1$ - phase

   c) $G_2$ – phase d) Telophase

5. Tall mutant of Dwarf Cavendish is:

   a) Grand Naine b) Robusta

   c) Gandevi d) Novaria

6. F1 crossed with one of its parents is known as:

   a) Test cross b) CRI

   c) Milking d) Grain filling

7. Critical stage of irrigation in wheat is:

   a) Tillering b) CRI

   c) Milking d) Grain filling

8. Dry farming is practiced where the amount of annual rainfall is:

   a) <750 mm b) >750 mm

   c) 750-1000 mm d) 1000 mm

9. Date palm is commercially propagated by:
   a) Suckers b) Runners
   c) Offshoots d) Stolen
10. In meiosis, crossing over takes place in:
    a) Metaphase b) Telophase
    c) Diplotene d) Pachytene
11. According to National Institute of Nutrition, the fruit having the highest anti- oxidant properties is:
    a) Orange b) Guava
    c) Pineapple d) Banana
12. The cropping system in which sowing of a crop is done when another crop is still standing in the field is known as:
    a) Alley cropping b) Relay cropping
    c) Mixed cropping d) Stripe cropping
13. Reclamation of acid soils is done by:
    a) Lime b) Gypsum
    c) Iron pyrites d) Rock phosphate
14. Gladiolus is commercially propagated by:
    a) Suckers b) Tubers
    c) Corms d) Cuttings
15. Botanical name of plum is:
    a) Prunus domestica b) Prunus avium
    c) Prunus cerasus d) Prunus persica
16. Quincunx system accommodates .... ......... times more plants than square system.
    a) 1 b) 1.5
    c) 2 d) 2
17. In order to avoid viral disease, a mid strain of pathogen is put into the genome of plant. This is known as:
    a) Immunization b) Sterilization
    c) Cross protection d) None of these
18. The regaining of growth by sugarcane crop stubbles is known as:
    a) Retooning b) Ratooning
    c) Suckering d) Tillering

19. India is the leading producer of:
    a) Mango b) Banana
    c) Sapota d) All of these
20. India is the highest producer of ...... in the world.
    a) Mango b) Banana
    c) Grapes d) Papaya
21. Spongy tissue is a serious problem in:
    a) Alphonso b) Dushehari
    c) Langra d) Pairi
22. Which cell organelle is also known as the "Power house of cell"?
    a) Ribosome b) Mitochondria
    c) Golgy body d) Nucleus
23. M-9 is dwarfing rootstock of:
    a) Mango b) Cherry
    c) Apple d) Pear
24. Forestro is a variety of:
    a) Cocoa b) Rubber
    c) Coconut d) Coffee
25. The degree of freedom in 2 x 3 factorial experiment with 2 replications will be:
    a) 21 b) 26
    c) 28 d) 30
26. Which of the following is a multiple fruit?
    a) Raspberry b) Custard apple
    c) Strawberry d) Pineapple
27. Botanical name of pummelo is:
    a) Citrus paradisi b) Citrus sinensis
    c) Citrus grandis d) Citrus aurantifolia
28. The indigenous variety of apple grown specifically in Kashmir is:
    a) Richard b) Red Gold
    c) Ambri d) Red Delicious

29. The head office of Coconut Development Board (CDB) is located at:
    a) Kochi b) Bengluru
    c) Thrissur d) Rahuri
30. Which of the following is a non-climacteric fruit?
    a) Banana b) Papaya
    c) Grape d) Mango
31. Which growth regulator is associated with stomatal regulation in plants?
    a) ABA b) Auxin
    c) Cytokinin d) GA3
32. Light reaction of photosynthesis occurs in:
    a) Stroma b) Grana
    c) Mitochondria d) Cytosol
33. The centromere of a chromosome is also known as:
    a) Chromomere b) Primary construction
    c) Telomere e) Secondary construction
34. ................ has the highest vitamin C content:
    a) Barbados cherry b) Aonla
    c) Guava d) Papaya
35. Arka Mridula is a variety of:
    a) Custard apple b) Pomegranate
    c) Guava d) Mango
36. Pineapple is propagated by:
    a) Suckers b) Slips
    c) Crowns d) All of these
37. Which of the following is a gynodioecious variety of papaya?
    a) Pusa Giant b) Coorg Honey Dew
    c) Pusa Dwarf d) Pusa Nanha
38. The following is a polyembryonic variety of mango:
    a) Fazli b) Gulab Khas
    c) Chandarkaran d) Keasr
39. Scab is a serious disease of:
    a) Mango b) Banana
    c) Apple d) Grape

40. Development of corky spots and fruit cracking in plum is due to the deficiency of:

    a) Molybdenum  b) Boron
    c) Calcium  d) Zinc

41. After fertilization, the fertilized flowers of coconut develop into mature nuts with in:

    a) 5-6 months  b) 7-8 months
    c) 9-10 months  d) 11-12 months

42. Tea mosquito bug is a serious pest of:

    a) Mango  b) Cashew nut
    c) Coconut  d) Banana

43. The following fruit is most suitable for making jam:

    a) Guava  b) Orange
    c) Banana  d) Pineapple

44. The most common rootstock used both mandarin and sweet orange is:

    a) Citron  b) Gajanimma
    c) Rough lemon  d) Sour orange

45. Commercially, cashew nut is not propagated by ..........:

    a) Epicotyl grafting  b) Soft wood grafting
    c) Air layering  d) Cleft grafting

46. In India, a major black pepper growing state is:

    a) West Bangal  b) Kerala
    c) Maharashtra  d) Odisha

47. Granulation is associated with:

    a) Sapota  b) Sweet orange
    c) Mango  d) Banana

48. Seedlessness in grape is induced by application of:

    a) GA3  b) Ethrel
    c) Kinetin  d) IAA

49. Which of the following is known as the 'King of Spices'?

    a) Cardamom  b) Black pepper
    c) Fenugreek  d) Cinnamon

50. A common PGR used for induction of rooting in cutting is:
    a) GA3 b) ABA
    c) IBA d) 2,4-D
51. ................ highly tolerant to low pH of soil:
    a) Cabbage b) Potato
    c) Onion d) Okra
52. Turmeric is propagated by:
    a) Tuber b) Rhizome
    c) Stem cutting d) Corms
53. Induction of male flowers in gynoecious lines of cucurbits is due to:
    a) Auxin b) Gibberellins
    c) Cytokinin d) ABA
54. Hot water treatment of seeds of cole crops to control internal seed borne disease is done at:
    a) 20°C for 30 minutes b) 30°C for 30 minutes
    c) 40°C for 15 minutes d) 50°C for 30 minutes
55. ......is not a dioecious vegetable:
    a) Sponge gourd b) Asparagus
    c) Pointed gourd d) Ivy gourd
56. The ploidy level and chromosome number of potato is:
    a) $2n = 4x = 48$ b) $2n = 2x = 48$
    c) $2n = 2x = 32$ d) $2n = 4x = 32$
57. Which of the following vegetables is autogamous:
    a) Radish b) Onion
    c) Palak d) Tomato
58. ................ is native of India:
    a) Pea b) Okra
    c) Brinjal d) Cowpea
59. Orchid seeds are devoid of:
    a) Seed coat b) Cotyledons
    c) Endosperm d) Embryo

60. Blossom end rot in tomato is caused due to the deficiency of:

a) Zinc b) Calcium

c) Boron d) Molybdenum

61. Most of the late cultivars of cabbage grown in India are:

a) Round headed b) Flat headed

c) Pointed headed d) Oblong type

62. Appearance of male female and hermaphrodite flowers in the same plant is:

a) Andromonoecious b) Gynomonoecious

c) Trimonoecious d) Sub androecious

63. An important antibiotic used for control of plant bacterial disease is:

a) Carbendazime b) Streptomycin sulphate

c) Sulphex d) Mancozeb

64. Which one of the following is used as non-selective herbicide?

a) Fluchloralin b) Pendimethalin

c) Paraquat d) Oxyflurofen

65. Head to seed method of seed production is followed in:

a) Cauliflower b) Onion

c) Cabbage d) Potato

66. The characteristic pungent flavour in radish is due to the presence of:

a) Allylpropyl disulphide b) Isothiocynate

c) Diallyl disulphide d) Allin

67. 'Baradari' is a canopied structure with .......... doors:

a) 6 b) 4

c) 12 d) 8

68. The largest cutflower producing state in India (2009-10 NHB data base) is:

a) Karnataka b) Andhra Pradesh

c) Maharashtra d) West Bengal

69. Chrysanthemum is popularly known as:

a) Japanese queen b) Queen of the East

c) Queen of China d) Winter queen

70. French marigold is a:

a) Diploid species b) Tetraploid species

c) Triploid species d) Hexaploid species

71. Which one is not a flowering annual?

a) Sweet pea b) Ipomoea

c) Kochia d) Calendula

72. Orchid species which grow on ground are known as?

a) Epiphyte b) Terrestrial

c) Lithophyte d) Saprophyte

73. Out of the total export of floricultural products from India, dry flowers constitute:

a) 20% b) 50%

c) 70% d) 30%

74. Inflorescence of gladiolus is:

a) Panicle b) Spadix

c) Spike d) Raceme

75. The botanical name of Mexican marigold is:

a) Tagetes patula b) Tagetes erecta

c) Tagetes tenuifolia d) Tagetes manuta

76. Which is not included in Japanese garden?

a) Hill garden b) Terrace garden

c) Flat garden d) Tea garden

77. Dianthus caryophyllus is the botanical name of:

a) Sweet William b) Dianthus

c) Carnation d) Corn flower

78. Gladiolus is propagated by:

a) Bulbs b) Corms

c) Tubers d) Rhizomes

79. The following flower is a rich source of xanthophyll pigment:

a) Chrysanthemum b) Marigold

c) China aster d) Dahlia

80. Gladiator is a variety of:
    a) Gladiolus b) Rose
    c) Carnation d) Chrysanthemum
81. ........ is an ethylene absorbing compound used for improving shelf life of cut flowers:
    a) Silver thiosulphate b) 8 HQC
    c) Boric acid d) ABA
82. The following shrub is grown for its ornamental leafy sepals:
    a) Calliandra b) Thunbergia
    c) Mussendra d) Bougainvillea
83. Enzyme responsible for converting protopectin into pectin is:
    a) Pectinase b) Proto pectinase
    c) Amylase d) Protease
84. The most suitable packaging material for cut flowers is:
    a) Ply-board boxes b) Plastic boxes
    c) Corrugated fibre board boxes d) Wooden boxes
85. ...... is the cheapest method of preservation.
    a) Canning b) Drying
    c) Fermentation d) Freezing
86. Which amino acid is the precursor of ethylene production in fruit tissue?
    a) Methionine b) Cysteine
    c) Glutamine d) Asparagine
87. Waxing helps in increasing the storage life of fresh fruits by:
    a) Reducing evaporation
    b) Reducing respiration
    c) Reducing evaporation and respiration
    d) Arresting microbial growth
88. Gherkins are extensively used for:
    a) Canning b) Pickling
    c) Salad preparation d) Dehydration
89. Pink flesh colour with reddish tinge of ripe watermelon is due to .........:
    a) Lycopene b) Anthocyanin
    c) Lycopene and anthocyanin d) Pelargonidin

90. Curcumin is extracted from:
    a) Turmeric b) Chilli
    c) Coriander d) Coloured capsicum
91. In most parts of India, pruning in rose is usually done during:
    a) October - November b) July - August
    c) August - September d) December - January
92. Pectin content is measured by:
    a) Refracto meter b) Jelly meter
    c) Jel meter d) Hydro meter
93. …….. is a leading mango growing state in India. :
    a) Karnataka b) Tamil Nadu
    c) Uttar Pradesh d) Punjab
94. Blanching helps in:
    a) Exhausting b) Fermentation
    c) Color fixing d) Powdering
95. Filler tree is accommodated in:
    a) Quincunx b) Hexagonal
    c) Contour d) Rectangular
96. Ally propyl disulphide is found in:
    a) Tomato b) Chilli
    c) Onion d) Brinjal
97. A suitable and easily available species for Bonsai:
    a) Rain tree b) Spathodia
    c) Copper pod tree d) Banyan tree
98. Seedlessness in banana is due to:
    a) Pseudospermocarpy b) Stenospermocarpy
    c) Vegetative parthenocarpy d) None of the above
99. To remove the field heat, fruits are dipped in:
    a) Liquid Ammonia b) Brine
    c) Water d) Dip oil
100. Taste, aroma and feel are referred to as:
    a) Astringency b) Rancidity
    c) Flavour d) Viscosity

101. The ideal plant for topiary:

a) Tecoma b) Clerodendron

c) Hemelia d) Hibiscus

102. The most important component of English garden is:

a) Border b) Rockery

c) Lawn d) Annuals

103. Adventitious embryony or nucellar embryony is found in:

a) Sapota b) Sweet Orange

c) Banana d) Ber

104. the commonly used rootstock in sapota is:

a) Rough lemon b) Bapakai

c) Rayan d) All of these

105. Exposure of potato tuber to sunlight causes:

a) Black heart b) Hollow heart

c) Browning d) Greening

106. The attached method of grafting is also known as:

a) Wedge grafting b) Approach grafting

c) Cleft grafting d) Saddle grafting

107. Little leaf in brinjal is caused due to:

a) Fugus b) Bacteria

c) Virus d) Mycoplasma

108. The word "Hortus" is derived from:

a) French b) Greek

c) Latin d) Persian

109. Degreening in sweet orange is done by using:

a) Ethylene b) IAA

c) IBA d) CCC

110. The economic part 'Head' is found in which of the following crop?

a) Cabbage b) Cauliflower

c) Knol - khol d) Brussels sprout

111. A common variety of banana grown by tissue culture is:

a) G-9 b) Rasthali

c) Poovan d) Metli

112. Hen and chicken is related to:
   a) Brinjal b) Grapes
   c) Tomato d) Mango
113. Which of the following crops belongs to the family of mango?
   a) Walnut b) Date plam
   c) Pecan nut d) Cashew nut
114. Which vegetable crop is commonly known as multivitamin green?
   a) Curry leaf b) Amaranthus
   c) Chekurmanis d) Methi
115. Which of the following is a leading export commodity of India?
   a) Onion b) Potato
   c) Garlic d) Cabbage
116. Export variety of pomegranate from India:
   a) Ganesh b) Keshar
   c) Jyoti d) Jalore Seedless
117. Tomato leaf curl is transmitted by:
   a) Thrips b) Aphid
   c) White fly d) Mites
118. The inflorescence of mango is:
   a) Panicle b) Raceme
   c) Umbel d) Catkin
119. Pisum sativum is the botanical name of:
   a) Cow pea b) French bean
   c) Cluster bean d) Garden pea
120. For longer shelf life, cushioning material is impregnated with:
   a) Ethrel b) Potassium permanganate
   c) Calcium carbonate d) Benzoic acid
121. Specific gravity indicates:
   a) Texture b) Firmness
   c) Maturity d) Flavour
122. ICARDA is situated at:
   a) Tokyo b) Syria
   c) Tasmania d) Turkmenistan

123. Banana variety used for cooking:

a) Poovan b) Monthan

c) Lal Kela d) Bombay Green

124. In which of the following fruit crops can nucellar seedlings be used to raise true-to- type plants?

a) Sapota b) Guava

c) Khirni d) Kagzi lime

125. Mango hybrid Mallika is a product of cross between:

a) Neelam x Dushehari b) Dushehari x Neelam

c) Alphonso x Neelam d) Neelam x Alphonso

126. The male sterility used in Ch-1 chilli hybrid development is:

a) Genetic male sterility b) Cytoplasmic male sterility

c) Cytoplasmic Genetic male sterility d) Functional male sterility

127. Multigerm seed is found in case of:

a) Carrot b) Radish

c) Turnip d) Beetroot

128. Which of the following fruit is the best for jelly making?

a) Mango b) Orange

c) Papaya d) Guava

129. To modify sex ratio is cucurbits, growth regulator is applied at:

a) 2-4, true leaf stage b) 4-6 true leaf stage

c) Flowering stage d) Fruiting stage

130. ............. is a dwarf mango hybrid:

a) Neelkiran b) Neeleshan

c) Neeluddin d) Amrapali

131. ...... is a serious disease in pomegranate threatening the area of expansion:

a) Bacterial wilt b) Leaf spot

c) Die back d) Bacterial blight

132. IIVR is located at:

a) New Delhi b) Lucknow

c) Varanasi d) Karnal

133. India's contribution in the world vegetable production is:

a) 3.38% b) 13.38%

c) 23.38% d) 33.38%

134. Juices which are served directly are known as:

a) Squash b) Cordial

c) RTS d) Marmalade

135. Frozen material needs to be:

a) Rehydrated b) Fumigated

c) Thawed d) Pasteurized

136. Lacquering means:

a) Lack of curing b) Lack of coating

c) Enamel coating d) Bursting

137. An export variety of apple from India is:

a) Newton Wonder b) Golden Delicious

c) Golden Russet d) Rome Beauty

138. .... is a very good specimen tree that does not have any shade effect on lawn is:

a) Rain tree b) Copper pod tree

c) Christmas tree d) Banyan tree

139. Softening of fruit is due to:

a) Loss of sugar b) Increase in acidity

c) Degradation of pectin d) Fungal infection

140. Enzyme present in dried latex of papaya is:

a) Protease b) Papain

c) Polygalacturonase d) Pectin esterase

141. ..... is a popular variety of loose skinned mandarin:

a) Kinnow mandarin b) Khasi mandarin

c) Coorg mandarin d) Nagpur mandarin

142. The common growth regulator for grape berry thinning and elongation is:

a) NAA b) Ethrel

c) BA d) GA3

143. Botanically, the fruit of pineapple is:

a) Berry b) Drupe

c) Sorosis d) Pome

144. ......... is an often cross pollinated vegetable is:

a) Tomato b) Onion

c) Okra d) Peas

145. Which of the following instruments is used for measuring leaf area?

a) Alanometer b) Anemometer

c) Porometer d) Barometer

146. maximum number of fruit setting in brinjal takes place .....type of flower.

a) Long style b) Medium style

c) Short style d) All of these

147. Seed plot technique has been developed for:

a) Onion b) Potato

c) Cabbage d) Radish

148. Chrysanthemum is a:

a) Short day plant b) Day neutral plant

c) Long day plant d) None of these

149. CITH is located at:

a) Jammu b) Srinagar

c) Bengaluru d) Shimla

150. Which one of the following is not a primary preservative?

a) Poassium metabisulphite b) Sugar

c) Oil d) Salt

151. Match the following:

i) Pulsing a) 8-HQC

ii) Herbicide b) Cycocel

iii) Biocide c) Sucrose

iv) Ripening hormone d) Glyphosate

v) Growth retardant e) Ethylene

152. Match the following:

i) Rose a) Bitter pit

ii) Cauliflower b) Cat face

| | | | |
|---|---|---|---|
| iii) | Mango | c) | Bent neck |
| iv) | Apple | d) | Riceyness |
| v) | Tomato | e) | Black tip |

153. Match the following:

| | | | |
|---|---|---|---|
| i) | Chrysanthemum | a) | Rhizome |
| ii) | Potato | b) | Corm |
| iii) | Canna | c) | Tuberous root |
| iv) | Sweet potato | d) | Suckers |
| v) | Freesia | e) | Stem tubers |

154. Match the following:

| | | | |
|---|---|---|---|
| i) | Whiptail in cauliflower | a) | Potassium |
| ii) | Interveinal chlorosis of young leaves | b) | Copper |
| iii) | Gum pockets at nodes of twigs | c) | Magnesium |
| iv) | Marginal and tip burning of leaves | d) | Molybdenum |
| v) | Interveinal chlorosis of older leaves | e) | Iron |

155. Match the following:

| | | | |
|---|---|---|---|
| i) | Stem borer | a) | Coconut |
| ii) | Shoot and fruit borer | b) | Citrus |
| iii) | Stem weevil | c) | Mango |
| iv) | Eryophide mite | d) | Brinjal |
| v) | Fruit sucking moth | e) | Banana |

156. Match the following:

| | | | |
|---|---|---|---|
| i) | Tristetza | a) | Fungus and bio-fertilizer |
| ii) | Pseudomonas | b) | Parasitoid wasp |
| iii) | VAM | c) | Fungus and bio-fungicide |
| iv) | Bracon hebetor | d) | Virus |
| v) | Trichoderma viride | e) | Bacterium |

157. Match the following:

| | | | |
|---|---|---|---|
| i) | Scarification | a) | Rooting inhibitor |
| ii) | Stratifications | b) | Sulphuric acid |

| | | | |
|---|---|---|---|
| iii) | Rhizocaline | c) | Moist chilling |
| iv) | TTC | d) | Chow - chow |
| v) | Vivipary | e) | Seed viability test |

158. Match the following:

| | | | |
|---|---|---|---|
| i) | Live hedge | a) | Citrus |
| ii) | Wind break | b) | Duranta plumeri |
| iii) | Viticulture | c) | Daincha |
| iv) | Citriculture | d) | Greveillea robusta |
| v) | Green manuring | e) | Grapes |

159. Match the following:

| | | | |
|---|---|---|---|
| i) | Asparagus | a) | Amaryllidaceae |
| ii) | Lettuce | b) | Chenopodiaceae |
| iii) | Colocasia | c) | Araceae |
| iv) | Leek | d) | Compositeae |
| v) | Beetroot | e) | Liliaceae |

160. Match the following:

| | | | |
|---|---|---|---|
| i) | 2,4-D | a) | Weedicide |
| ii) | Glyphosate | b) | Herbicide |
| iii) | Ethylene | c) | Ripening hormone |
| iv) | ABA | d) | Stomatal |
| v) | Azotobacter | e) | Free living bacteria |

## Answer Key

| | | | | | | | | | | | | | |
|---|---|---|---|---|---|---|---|---|---|---|---|---|---|
| 1. | (a) | 2. | (c) | 3. | (c) | 4. | (a) | 5. | (a) | 6. | (b) | 7. | (b) |
| 8. | (a) | 9. | (c) | 10. | (d) | 11. | (b) | 12. | (b) | 13. | (a) | 14. | (c) |
| 15. | (a) | 16. | (b) | 17. | (c) | 18. | (b) | 19. | (d) | 20. | (c) | 21. | (a) |
| 22. | (b) | 23. | (c) | 24. | (a) | 25. | (a) | 26. | (d) | 27. | (c) | 28. | (c) |
| 29. | (a) | 30. | (c) | 31. | (a) | 32. | (b) | 33. | (b) | 34. | (a) | 35. | (c) |
| 36. | (d) | 37. | (b) | 38. | (c) | 39. | (c) | 40. | (b) | 41. | (d) | 42. | (b) |
| 43. | (d) | 44. | (c) | 45. | (c) | 46. | (b) | 47. | (b) | 48. | (a) | 49. | (b) |
| 50. | (c) | 51. | (b) | 52. | (b) | 53. | (b) | 54. | (d) | 55. | (a) | 56. | (a) |
| 57. | (d) | 58. | (c) | 59. | (c) | 60. | (b) | 61. | (b) | 62. | (c) | 63. | (b) |
| 64. | (c) | 65. | (c) | 66. | (b) | 67. | (c) | 68. | (d) | 69. | (b) | 70. | (b) |
| 71. | (c) | 72. | (b) | 73. | (b) | 74. | (c) | 75. | (b) | 76. | (b) | 77. | (c) |

| | | | | | | | | | | | | | |
|---|---|---|---|---|---|---|---|---|---|---|---|---|---|
| 78. | (b) | 79. | (b) | 80. | (b) | 81. | (a) | 82. | (c) | 83. | (b) | 84. | (c) |
| 85. | (b) | 86. | (a) | 87. | (b) | 88. | (b) | 89. | (c) | 90. | (a) | 91. | (a) |
| 92. | (c) | 93. | (c) | 94. | (c) | 95. | (a) | 96. | (c) | 97. | (d) | 98. | (c) |
| 99. | (c) | 100. | (c) | 101. | (b) | 102. | (c) | 103. | (b) | 104. | (c) | 105. | (d) |
| 106. | (b) | 107. | (d) | 108. | (c) | 109. | (a) | 110. | (a) | 111. | (a) | 112. | (b) |
| 113. | (d) | 114. | (c) | 115. | (a) | 116. | (a) | 117. | (c) | 118. | (a) | 119. | (d) |
| 120. | (b) | 121. | (c) | 122. | (b) | 123. | (b) | 124. | (d) | 125. | (a) | 126. | (a) |
| 127. | (d) | 128. | (d) | 129. | (a) | 130. | (d) | 131. | (d) | 132. | (c) | 133. | (b) |
| 134. | (c) | 135. | (c) | 136. | (c) | 137. | (b) | 138. | (c) | 139. | (c) | 140. | (b) |
| 141. | (d) | 142. | (d) | 143. | (c) | 144. | (c) | 145. | (a) | 146. | (a) | 147. | (b) |
| 148. | (a) | 149. | (b) | 150. | (a) | | | | | | | | |

| Q. | 151. | 152. | 153. | 154. | 155. | 156. | 157. | 158. | 159. | 160. |
|---|---|---|---|---|---|---|---|---|---|---|
| i) | (c) | (c) | (d) | (d) | (c) | (d) | (b) | (b) | (e) | (a) |
| ii) | (d) | (d) | (e) | (e) | (d) | (e) | (c) | (d) | (d) | (b) |
| iii) | (a) | (e) | (a) | (b) | (e) | (a) | (a) | (e) | (c) | (c) |
| iv) | (e) | (a) | (c) | (a) | (a) | (b) | (e) | (a) | (a) | (d) |
| v) | (b) | (b) | (b) | (c) | (b) | (c) | (d) | (c) | (b) | (e) |

# 4

# ICAR – JRF Horticulture Exam – 2013

1. The total annual rainfall of India is about:

   a) 1190 mm  b) 1250 mm
   c) 1050 mm  d) 1350 mm

2. Net cultivated area of India is:

   a) 142 mha  b) 150 mha
   c) 140 mha  d) 152 mha

3. Amrapali is a cross between:

   a) Neelum x Dushehari  b) Dushehari x alphonso
   c) Ratna x Alphonso  d) Dushehari x Neelum

4. Bengal famine was caused by:

   a) Phytopthora infestance  b) Helminthosporium oryzaae
   c) Hemelia vestratrix  d) Colletotrichum falcatum

5. In meiosis, crossing over takes place in:

   a) Diplotene  b) Zygotene
   c) Pachytene  d) Leptotene

6. Mutation which kills less than 50% population is known as:

   a) Lethal  b) Sub-lethal
   c) Sub-vital  d) Vital

7. A dwarf mutant variety of papaya is:

   a) Pusa Dwarf  b) Pusa Nanha
   c) Pusa Giant  d) Pusa Majesty

8. Khaira disease of rice is caused by:

   a) Fungus  b) Bacteria
   c) Virus  d) None of these

9. The most ancient type of vegetable garden is:
   a) Kitchen garden b) Market garden
   c) Truck garden d) Floating garden
10. Which of the following is a single seeded nut?
   a) Cashew nut b) Walnut
   c) Litchi d) Peacan nut
11. Seedlessness in grape is due to:
   a) Vegetative parthenocarpy b) Stimulative parthenocarpy
   c) Stenospermocarpy d) Embryo abortion
12. Which of the following crop can be grown in acidic soil?
   a) Water melon b) Musk melon
   c) Round melon d) Long melon
13. Red colour of carrot is due to the presence of:
   a) Anthocyanin b) Lycopene
   c) Pelargonidin d) Delphinidin
14. Gulkand is prepared by mixing rose petals and sugar in a ratio of:
   a) 1 : 1 b) 1 : 2
   c) 1 : 3 d) 2 : 1
15. Botanical name of peach is:
   a) Prunus domestica b) Prunus avium
   c) Prunus cerasus d) Prunus persica
16. Pusa Tara is a cultivar of:
   a) Coreopsis b) Hibiscus
   c) Tuberose d) Gladiolus
17. Which of the following is a sexually sterile diploid?
   a) Leek b) Garlic
   c) Onion d) None of these
18. The storage temperature of banana is:
   a) 10°C b) 13°C
   c) 15°C d) 8°C
19. Xeriscaping refers to:
   a) Use of xerophytic plants for indoor decoration
   b) Landscaping in arid areas
   c) Soil conservation measures
   d) All of these

20. Economic part of isabgol is:
    a) Seeds  b) Husk
    c) Both seeds and husk  d) None of these
21. Offsets are very common for the prorogation of:
    a) Aloe  b) Pandanus
    c) Agave  d) All of these
22. The seeds distributed to farmers for cultivation are referred to as :
    a) Breeder seed  b) Foundation seed
    c) Nucleus seed  d) Certified seed
23. Which of the following is rich in iron?
    a) Musk melon  b) Water melon
    c) Bitter gourd  d) Bottle gourd
24. Which of the following states occupies the largest area in chilli cultivation?
    a) Gujarat  b) Maharashtra
    c) Andhra Pradesh  d) Karnataka
25. Caprification is related to which of the following crops:
    a) Fig  b) Avocado
    c) Loquat  d) Persimmon
26. Fire blight of pear is caused by:
    a) Psedomonas syringeae  b) Erwinia crotovora
    c) Erwinia amylovora  d) Erwinia herbicola
27. Cricket Ball is a variety of:
    a) Sapota  b) Orange
    c) Pear  d) Bael
28. Which of the following cucurbits contains the highest amount of vitamin A:
    a) Pumpkin  b) Summer squash
    c) Winter squash  d) Bottle gourd
29. The permissible limit of added colour in food products is:
    a) 100 ppm  b) 200 ppm
    c) 250 ppm  d) 300 ppm
30. In nature, most of the cultivated banana belongs to .......group:
    a) Diploid  b) Triploid
    c) Tetraploid  d) Hexaploid

31. Which breeding method is used for the purification of existing variety?
    - a) Mass selection
    - b) Pedigree method
    - c) Bulk method
    - d) Recurrent selection

32. Maximum inbreeding depression as a result of selfing is observed in which of the following crop:
    - a) Cucumber
    - b) Carrot
    - c) Cabbage
    - d) Radish

33. Regularity in bearing in mango can be induced to some extent with the application of:
    - a) Paclobutrazol
    - b) Auxin
    - c) Ethephon
    - d) ABA

34. Which disease mostly occurs in high density planting in banana?
    - a) Sigatoka leaf spot
    - b) Bunchy top
    - c) Panama wilt
    - d) Finger tip

35. The concentration of vinegar in preservative should not be more than:
    - a) 5%
    - b) 4%
    - c) 3%
    - d) 2%

36. Jam marmalade is prepared from the pulp of which of the following fruits?
    - a) Mango
    - b) Guava
    - c) Citrus
    - d) Pomegranate

37. Which kind of cucumber hybrids are most suitable for cultivation under protected environment?
    - a) Monoecious
    - b) Parthenocarpic
    - c) Andromonoecious
    - d) Gynoecious

38. In guava, 'Mrig Bahar' season bears flowering during which months?
    - a) November - December
    - b) February – March
    - c) June - July
    - d) October – November

39. Which of the following is propagated by viviparous single seed fruits?
    - a) Momordica charantia
    - b) Luffa acutangula
    - c) Cucumis melo
    - d) Sechium edule

40. The most concentrated fertilizer used for nutrient supply is:
    - a) Urea
    - b) DAP
    - c) Anhydrous Ammonia
    - d) SSP

41. Which of the following is a non spur apple variety?
    a) Starkrimson b) Red Chief
    c) Vance Delicious d) Oregon spur
42. Which of the following bears climacteric fruits?
    a) Water Melon b) Long Melon
    c) Musk Melon d) Round Melon
43. Spongy tissue is a serious disorder hampering the export of which mango variety:
    a) Alphonso b) Dushehari
    c) Neelum d) Langra
44. The best planting material for pineapple is:
    a) Slips b) Crown
    c) Suckers d) Corms
45. Total area under vegetable cultivation in India is:
    a) 6.28 million ha b) 7.98 million ha
    c) 8.49 million ha d) 9.22 million ha
46. Which of the following systems of training grapes has high cost benefit ratio and is followed commercially?
    a) Bower b) Head
    c) Kniffin d) Trellis
47. Which group of vegetables is the most suitable for river bed cultivation?
    a) Cole crops b) Root crops
    c) Bulb crops d) Cucurbits
48. Green leafy vegetables are rich in:
    a) Phosphorus b) Sodium
    c) Iron d) Copper
49. Most common sex expression in cucurbits is:
    a) Monoecism b) Dioecism
    c) Andromonoecism d) Gynoecism
50. Delayed harvesting in radish may cause:
    a) Forking b) Pithiness
    c) Deformed roots d) Cavity spot

51. Micro-grafting is used to produce plants that are free from:
    a) Viruses  b) Bacteria
    c) Fungi  d) Nematodes
52. Most widely used plant growth regulator used in viticulture for berry elongation is:
    a) GA3  b) NAA
    c) 2,4-D  d) IAA
53. Guava grows mostly on:
    a) One year old shoots  b) Spurs
    c) Current season growth  d) Very old shoots
54. Which of the following is not a pome fruit?
    a) Apple  b) Pear
    c) Peach  d) Loquat
55. Ikebana is a kind of:
    a) Pot plant arrangement  b) Cultivation without soil
    c) Flower arrangement  d) Bonsai type
56. Indian cauliflower is mainly:
    a) Self - compatible  b) Self - incompatible
    c) Self – sterile  d) Cross – incompatible
57. Which country is the largest exporter of cut flower in the world?
    a) USA  b) Russia
    c) Japan  d) The Netherland
58. Apomixis in citrus is of the ...............type:
    a) Polyembryony  d) Recurrent
    c) Vegetative  d) Non – recurrent
59. Which of the following cucurbits is dioecious in nature?
    a) Round gourd  b) Bottle gourd
    c) Pointed gourd  d) Bitter gourd
60. Bitterness in tea is due to:
    a) Caffeine  b) Sinigrin
    c) Amygladine  d) Naringin

61. According to FPO, maximum amount of potassium metabisulphite allowed in juice is:
    a) 500 ppm
    b) 600 ppm
    c) 700 ppm
    d) 800 ppm
62. Which of the following has large sized white and fragrant flowers?
    a) Jacaranda
    b) Magnolia
    c) Flame of the forest
    d) Nyctanthes
63. Cross pollination in spinach is facilitated by:
    a) Insects
    b) Wind
    c) Water
    d) Birds
64. Which is the finest mandarin variety in the world?
    a) Coorg
    b) Nagpur
    c) Satsuma
    d) Khasi
65. Which commercial propagating method is used in Rose?
    a) T budding
    b) Tongue grafting
    c) Layering
    d) Seed
66. Which principle is to be adopted in the garden to show a continuous effect?
    a) Balance
    b) Harmony
    c) Accent
    d) Proportion
67. The training system followed in pomegranate is:
    a) Open centre
    b) Central leader
    c) Multistem
    d) Spindle bush
68. Pasteurization of fruit and vegetable juice is done at:
    a) 85-90°C for 30 minutes
    b) 105-110°C for 30 minutes
    c) 45-50°C for 30 minutes
    d) 65-70°C for 30 minutes
69. What kind of male sterility can be utilized for hybrid seed production in cole crops?
    a) GMS
    b) FMS
    c) CMS
    d) CGMS
70. Pheromone traps are used for monitoring pest population of:
    a) Aphids
    b) Fruit fly
    c) White fly
    d) Jassids

71. Name the cucurbit which is used as candied or preserved:
    a) Long gourd b) Musk melon
    c) Water melon d) Ash gourd
72. Which of the following is a shade loving pot plant?
    a) Anthurium b) Dieffenbachia
    c) Croton d) Viola
73. Which part of clove used for consumption is:
    a) Seed b) Bark
    c) Flower d) Unopened bud
74. Edible part of asparagus is:
    a) Root b) Tender shoots
    c) Fruits d) Flowers
75. Propagation of large cardamom is done by:
    a) Seed b) Rhizome
    c) Cutting d) Air layering
76. Which medicinal plant has sedative property?
    a) Ashwagandha b) Sarpagandha
    c) Isabgol d) Periwinkle
77. Pungency in garlic is due to:
    a) Allyl propyl disulphide b) Dimethyl disulphide
    c) Diallyl disulphide d) Isothiocynate
78. Black heart in potato is the result of:
    a) Moisture deficiency b) Improper storage conditions
    c) Low temperature d) Excessive N2
79. Which one of the following plant nutrients is useful in increasing resistance to diseases and insects pests?
    a) Calcium b) Phosphors
    c) Nitrogen d) Potassium
80. Which variety of banana requires propping?
    a) Dwarf Cavendish b) Robusta
    c) Grand Naine d) Poovan

81. Name the Indian state where garden pea can be grown as an offseason crop during summer?

a) Punjab b) Maharashtra
c) Himachal Pradesh d) Karnataka

82. ICMR recommendations for consumption of vegetable/capita/day by an adult is around:

a) 150g b) 250g
c) 300g d) 400g

83. An ideal fruit for making jelly should be rich in:

a) Pectin and sugars b) Acids and proteins
c) Sugars and acids d) Pectin and acids

84. Oleoresin is extracted from:

a) Onion b) Garlic
c) Chilli d) Fenugreek

85. The Permanent method of preservation of fruits and vegetables is:

a) Pasteurization b) Blanching
c) Canning and Bolting d) Exclusion of moisture

86. The quickest method to grow a lawn is through:

a) Seeding b) Dibbling
c) Turfing d) Plastering

87. Which is a C-4 plant?

a) Amaranthus b) Sweet potato
c) Okra d) Garden pea

88. Which variety of apple has long keeping quality?

a) Rymer b) Red Delicious
c) Ambri d) Golden Delicious

89. Name the nutrient that is highly mobile in the plants:

a) N b) S
c) Mg d) B

90. Which type of garden is called "Nature in Miniature":

a) Japanese b) Mughal
c) Buddhist d) English

91. Which state of India is the leading tea producer?
    a) Assam  b) Himachal Pradesh
    c) West Bengal  d) Tamil Nadu
92. Most expensive spice in the world is:
    a) Large Cardamom  b) Cloves
    c) Saffron  d) Nutmeg
93. Bacterial diseases are controlled by application of:
    a) Viricides  b) Fungicides
    c) Antibiotics  d) Kerathane
94. Cider, a fermented vine, is prepared from:
    a) Grapes  b) Cashew nut
    c) Pear  d) Apple
95. A fruit crop which is tolerant to salinity is:
    a) Mango  b) Orange
    c) Ber  d) Apple
96. Clonal rootstock of apple can be easily propagated through:
    a) Cutting  b) Stooling
    c) Air layering  d) Tongue grafting
97. GM brinjal has been developed to manage:
    a) Little leaf  b) Shoot and fruit borer
    c) Hadda beetle  d) Spider mites
98. Flowers are borne terminally in:
    a) Mango  b) Jackfruit
    c) Coffee  d) Cocoa
99. Amrapali is a cross between:
    a) Neelum x Dushehari  b) Dushehari x Alphonso
    c) Ratna x Alphonso  d) Dushehari x Neelum
100. Origin of pineapple is:
    a) India  b) Brazil
    c) China  d) Australia
101. Ullal series is related to cultivars of:
    a) Tea  b) Coffee
    c) Cashew  d) Cocoa

102. Laksha Ganga is an improved hybrid of:

| | | | |
|---|---|---|---|
| a) | Arecanut | b) | Coconut |
| c) | Pecan nut | d) | Walnut |

103. The bitterness/acridity in colocasia corm is due to:

| | | | |
|---|---|---|---|
| a) | Calcium oxalate | b) | Calcium carbonate |
| c) | Calcium chloride | d) | Potassium oxalate |

104. Pendimethalin and butachlor are:

| | | | |
|---|---|---|---|
| a) | Post-emergent weedicide | b) | Pre-emergent weedicide |
| c) | Early post emergent weedicide | d) | Emergent weedicide |

105. Preserving rose with sugar is known as:

| | | | |
|---|---|---|---|
| a) | Pankuri | b) | Gulkand |
| c) | Concrete | d) | Absolute of rose |

106. Boron deficiency in tomato causes:

| | | | |
|---|---|---|---|
| a) | Browning | b) | Fruit cracking |
| c) | Puffiness | d) | Cat facing |

107. The ideal month for pruning Jasminium multiflorum is:

| | | | |
|---|---|---|---|
| a) | September - October | b) | March - April |
| c) | December - January | d) | June – July |

108. Bending of shoots is practiced in:

| | | | |
|---|---|---|---|
| a) | Carnation | b) | Rose under greenhouse |
| c) | Chrysanthemum | d) | Dahlia |

109. Boron deficiency in carnation causes:

| | | | |
|---|---|---|---|
| a) | Bent neck | b) | Calyx splitting |
| c) | Stem splitting | d) | Bull head |

110. Which family does litchi belong to?:

| | | | |
|---|---|---|---|
| a) | Rosaceae | b) | Spindaceae |
| c) | Anacardiaceae | d) | Myrtaceae |

111. The best fruit recommended for diabetes is:

| | | | |
|---|---|---|---|
| a) | Mango | b) | Avocado |
| c) | Sitaphal | d) | Jamun |

112. The family of gladiolus is:

| | | | |
|---|---|---|---|
| a) | Caryophyllaceae | b) | Pinaceae |
| c) | Iridaceae | d) | Compositae |

113. The basic chromosome number of rose is:
    a) 8 b) 6
    c) 7 d) 9
114. Pungency in chilli is caused by:
    a) Capsaicin b) Capsnithin
    c) Caprocin d) Capsicum
115. Bitter pit in apple is caused due to the deficiency of which element:
    a) Magnesium b) Sulphur
    c) Calcium d) Nitrogen
116. Chrysanthemum is a:
    a) Long day plant b) Short day plant
    c) Day neutral plan d) Photo insensitive
117. Water melon bud necrosis is caused by:
    a) Fungus b) Bacteria
    c) Virus d) Mycoplasma
118. The commercially grown kharif onion variety is:
    a) Pusa Red b) Arka Niketan
    c) N-53 d) Banglore Rose
119. The type of male sterility found in marigold is:
    a) Genetic b) Cytoplasmic
    c) Cytoplasmic genetic d) Self-incompatibility
120. The best criterion for determining maturity indices in citrus is:
    a) TSS content b) Sugar content
    c) TSS / acid ratio d) Acidity
121. Red lady is a variety of:
    a) Grape b) Banana
    c) Tangerine d) Papaya
122. Which one of the following is the cause of self-unfruitful in loquat?
    a) Gametophytic incompatibility b) Pollen sterility
    c) Ovule abortion d) Cross incompatibility
123. Which of the following is pome fruit?
    a) Cherry b) Almond
    c) Loquat d) Strawberry

124. Where was Prabhat peach developed?:
   a) PAU, Ludhiana  b) UHF, Nauni
   c) CCSHAU, Hisar  d) HETC, Saharanpur

125. A variety of apricot suitable for dry desert areas is:
   a) Shipley  b) Shakarpara
   c) Nuggets  d) Early Shipley

126. Major acid found in apple is:
   a) Malic acid  b) Lactic acid
   c) Mucic acid  d) Citamalic acid

127. Salt tolerant rootstock of mango is:
   a) 13/1  b) Mulgoa
   c) Cecil  d) Nikara

128. Albedo in citrus fruit is rich in:
   a) Protein  b) Oil
   c) Cellulose  d) Vitamin C

129. NRC on seed spices is situated at:
   a) New Delhi  b) Jodhpur
   c) Ajmer  d) Kasaragod

130. Which is the first synthetic auxin to be used as herbicide?
   a) 2,4-D  b) 2,4,5-T
   c) MCPA  d) TCDD

131. Intercropping of tomato with cabbage can reduce the damage done by the:
   a) Diamond back moth  b) Aphids
   c) Butterflies  d) Leafhoppers

132. The most appropriate time for pruning of rose in north India is:
   a) October - November  b) December - January
   c) March - April  d) August – September

133. Calyx-splitting is a physiological disorder of which crop:
   a) Carnation  b) Dahlia
   c) Chrysanthemum  d) Cauliflower

134. Which of the following grasses is not suitable for cooler regions?
   a) Kentucky blue grass  b) Red top grass
   c) Rye grass  d) Korean grass

135. Clear juice are obtained by treating with:
   a) Carrageenan  b) Gum bacteria
   c) Bentonite  d) Gelatin

136. The red pigment found in onion is:
   a) Anthocyanin  b) Lycopene
   c) Capstheum  d) Allium

137. Which microorganism is of serious concern in reduced oxygen packaging?
   a) E. coli  b) Penicillium spp.
   c) Bacillus spp.  d) Listeria monocytogens

138. Chemically pectin is :
   a) Fibrometer  b) Instron
   c) Hunter lab  d) Texturizer

140. Canna is propagated by:
   a) Bulb  b) Corm
   c) Rhizome  d) Tuber

141. The centre of attraction in a garden is known as:
   a) Divisional line  b) Focal point
   c) Axis  d) Sun bath

142. A series of arches joined together is known as:
   a) Mega arch  b) Dome
   c) Pergola  d) Topiary

143. The quickest method of developing a lawn is:
   a) Seeding  b) Dibbling
   c) Turf plastering  d) Turfing

144. Palmarosa grass gives us:
   a) Valuable fodder  b) Essential oil
   c) Vegetable oil  d) Kerosene oil

145. International Flower Auction Centre in India in located at:
   a) Bengaluru  b) New Delhi
   c) Chennai  d) West Bengal

146. The headquarter of PPV & FR Authority is located at:
   a) Hyderabad  b) New Delhi
   c) Trivandrum  d) Chennai

147. Carnation flowers are harvested at:

| | | | |
|---|---|---|---|
| a) | Bud stage | b) | Half open stage |
| c) | Paint brush stage | d) | Fully open stage |

148. R.R. Pal is a variety of:

| | | | |
|---|---|---|---|
| a) | Bougainvillea | b) | Gladiolus |
| c) | Amaranthus | d) | Pumpkin |

149. The variety of Chrysanthemum viz. 'Pusa Anmol' is a mutant of:

| | | | |
|---|---|---|---|
| a) | Ajay | b) | Usha |
| c) | Vijay | d) | Red Gold |

150. "Damping off" disease in nursery seedlings is caused by:

| | | | |
|---|---|---|---|
| a) | Fungus | b) | Bacteria |
| c) | Virus | d) | MLO |

151. Match the following:

| | | | |
|---|---|---|---|
| i) | Tree | a) | Begonia |
| ii) | Shrub | b) | Kochia |
| iii) | Climber | c) | Jacaranda |
| iv) | Pot plant | d) | Camellia |
| v) | Annual | e) | Cat's claw |

152. Match the following

| | | | |
|---|---|---|---|
| i) | Black tip of mango | a) | Mo |
| ii) | Die back of citrus | b) | B |
| iii) | Bronzing of guava | c) | Ca |
| iv) | Whiptail in cauliflower | d) | Zn |
| v) | Cavity spot in carrot | e) | Cu |

153. Match the following

| | | | |
|---|---|---|---|
| i) | CIPHET | a) | Anand (Gujarat) |
| ii) | NRCM | b) | Ajmer (Rajasthan) |
| iii) | NRCM & AP | c) | New Delhi |
| iv) | NRCSS | d) | Ludhiana (Punjab) |
| v) | NRCPB | e) | Solan (Himachal Pradesh) |

154. Match the following

| | | | |
|---|---|---|---|
| i) | Banana | a) | Catkin |
| ii) | Pomegranate | b) | Spadix |
| iii) | Tomato | c) | Umble |
| iv) | Onion | d) | Hypanthodium |
| v) | Cabbage | e) | Racemose |

155. Match the following

| | | | |
|---|---|---|---|
| i) | Chasme shahi | a) | Bangalore |
| ii) | Mandore garden | b) | New Delhi |
| iii) | Rock garden | c) | Lahore |
| iv) | Lal bagh garden | d) | Chandigarh |
| v) | Buddha jayanti garden | e) | Jodhpur |

156. Match the following

| | | | |
|---|---|---|---|
| i) | IISR | a) | Shimla |
| ii) | CAZRI | b) | Lucknow |
| iii) | CIMAP | c) | Jodhpur |
| iv) | CPRI | d) | Calicut |
| v) | CIAH | e) | Bikaner |

157. Match the following

| | | | |
|---|---|---|---|
| i) | Tuberose | a) | Corm |
| ii) | Carnation | b) | Bulb |
| iii) | Gladiolus | c) | Cutting |
| iv) | Marigold | d) | Seeds |
| v) | Rose | e) | Budding |

158. Match the following

| | | | |
|---|---|---|---|
| i) | Clerodendron | a) | Aristolochiaceae |
| ii) | Duck flower | b) | Verbenaceae |
| iii) | Statice | c) | Sea lavender |
| iv) | Railway creeper | d) | White flower |
| v) | Jasmine | e) | Purple flower |

159. Match the following:

| | | | |
|---|---|---|---|
| i) | Basic nutrient | a) | C |
| ii) | Primary nutrient | b) | N |
| iii) | Secondary nutrient | c) | S |
| iv) | Micro nutrient | d) | Zn |
| v) | Beneficial nutrient | e) | Si |

160. Match the following:

| | | | |
|---|---|---|---|
| i) | 2,4-D | a) | Weedicide |
| ii) | Glyphosate | b) | Herbicide |
| iii) | Ethylene | c) | Ripening hormone |
| iv) | ABA | d) | Stomatal regulation |
| v) | Azotobacter | e) | Free living bacteria |

## Answers Key

| | | | | | | | | | | | | | |
|---|---|---|---|---|---|---|---|---|---|---|---|---|---|
| 1. | (a) | 2. | (a) | 3. | (d) | 4. | (b) | 5. | (c) | 6. | (c) | 7. | (b) |
| 8. | (d) | 9. | (a) | 10. | (c) | 11. | (c) | 12. | (a) | 13. | (b) | 14. | (a) |
| 15. | (d) | 16. | (a) | 17. | (b) | 18. | (b) | 19. | (c) | 20. | (c) | 21. | (d) |
| 22. | (d) | 23. | (c) | 24. | (c) | 25. | (a) | 26. | (b) | 27. | (a) | 28. | (a) |
| 29. | (b) | 30. | (b) | 31. | (a) | 32. | (b) | 33. | (a) | 34. | (d) | 35. | (d) |
| 36. | (c) | 37. | (b) | 38. | (c) | 39. | (d) | 40. | (c) | 41. | (c) | 42. | (c) |
| 43. | (a) | 44. | (a) | 45. | (c) | 46. | (a) | 47. | (d) | 48. | (c) | 49. | (a) |
| 50. | (b) | 51. | (a) | 52. | (a) | 53. | (c) | 54. | (c) | 55. | (c) | 56. | (b) |
| 57. | (d) | 58. | (a) | 59. | (c) | 60. | (a) | 61. | (c) | 62. | (b) | 63. | (b) |
| 64. | (b) | 65. | (a) | 66. | (b) | 67. | (c) | 68. | (a) | 69. | (c) | 70. | (b) |
| 71. | (d) | 72. | (b) | 73. | (d) | 74. | (b) | 75. | (b) | 76. | (b) | 77. | (c) |
| 78. | (b) | 79. | (d) | 80. | (c) | 81. | (c) | 82. | (c) | 83. | (d) | 84. | (c) |
| 85. | (d) | 86. | (c) | 87. | (a) | 88. | (c) | 89. | (a) | 90. | (a) | 91. | (a) |
| 92. | (c) | 93. | (c) | 94. | (d) | 95. | (c) | 96. | (b) | 97. | (b) | 98. | (a) |
| 99. | (d) | 100. | (b) | 101. | (c) | 102. | (b) | 103. | (a) | 104. | (b) | 105. | (b) |
| 106. | (b) | 107. | (c) | 108. | (b) | 109. | (b) | 110. | (b) | 111. | (d) | 112. | (c) |
| 113. | (c) | 114. | (a) | 115. | (c) | 116. | (b) | 117. | (c) | 118. | (c) | 119. | (a) |
| 120. | (c) | 121. | (d) | 122. | (a) | 123. | (c) | 124. | (d) | 125. | (b) | 126. | (a) |
| 127. | (a) | 128. | (c) | 129. | (c) | 130. | (a) | 131. | (a) | 132. | (a) | 133. | (a) |
| 134. | (a) | 135. | (d) | 136. | (a) | 137. | (a) | 138. | (b) | 139. | (d) | 140. | (c) |
| 141. | (b) | 142. | (c) | 143. | (d) | 144. | (b) | 145. | (a) | 146. | (b) | 147. | (c) |
| 148. | (a) | 149. | (a) | 150. | (a) | | | | | | | | |

| Q. | 151. | 152. | 153. | 154. | 155. | 156. | 157. | 158. | 159. | 160. |
|---|---|---|---|---|---|---|---|---|---|---|
| i) | (e) | (b) | (d) | (b) | (c) | (d) | (b) | (b) | (a) | (a) |
| ii) | (c) | (e) | (e) | (d) | (e) | (c) | (c) | (a) | (b) | (b) |
| iii) | (d) | (d) | (a) | (e) | (d) | (b) | (a) | (c) | (c) | (c) |
| iv) | (a) | (a) | (b) | (c) | (a) | (a) | (d) | (e) | (d) | (d) |
| v) | (b) | (c) | (c) | (a) | (b) | (e) | (e) | (d) | (e) | (e) |

# 5

# ICAR – JRF Horticulture Exam – 2014

1. The first Indian scientist who won the 'World Food Prize' is:
   a) Sanjay Rajaram b) M.S. Swaminathan
   c) Vergese Kurien d) G.S. Khus
2. Seed rate of paddy (kg/ha) in SRI (System of Rice Intensification) method of cultivation is:
   a) 5-6 b) 8-10
   c) 12-14 d) 18-20
3. AVRDC is situated at:
   a) Phillipines b Taiwan
   c) Japan d) India
4. CIAH is located at:
   a) Srinagar b) Varanasi
   c) Jodhpur d) Bikaner
5. Lux is the unit of measurement of:
   a) Light intensity b) Light quality
   c) Light duration d) All of these
6. Enology is the study of:
   a) Monument b) Wines
   c) Drying of horticultural produce d) *In-vitro* propagation
7. Esmasculation is done in:
   a) Self-pollinated crops b) Cross pollinated crops
   c) Often cross-pollinated crops d) Self-incompatible crops
8. Perry is prepared from:
   a) Pear b) Peach
   c) Plum d) Persimmon

9. The fruit type of fig is:
   a) Amphisaraca  b) Sorosis
   c) Syconus  d) Balausta
10. Foundation seed is the progeny of:
   a) Breeder seed  b) Registered seed
   c) Certified seed  d) All of these
11. 'Flavr Savr' is the transgenic variety of:
   a) Summer squash  b) Papaya
   c) Brinjal  d) Tomato
12. Flavour includes:
   a) Taste  b) Aroma
   c) Both taste and aroma  d) Texture
13. Wet meat is a product of:
   a) Cashew nut  b) Coconut
   c) Arecanut  d) Walnut
14. Neera is a fermented product of:
   a) Coconut  b) Cashew nut
   c) Date palm  d) Palmyra palm
15. The flower which is known as the "Queen of East" is:
   a) Carnation  b) Chrysanthemum
   c) Rose  d) Gladiolus
16. Hexagonal system of planting accommodates % more plants than square system.
   a) 10  b) 15
   c) 20  d) 25
17. The botanical name of Jungal Jalebi (manila tamarind) is:
   a) *Pithecellobium dulce*  b) *Tamarindus indica*
   c) *Prosopis cineraria*  d) *Salvadora persica*
18. Contender is a variety of:
   a) Lime bean  b) French bean
   c) Winged bean  d) Faba bean

19. Tomato leaf curl virus is transmitted by:
    a) Aphids b) Jassids
    c) Fruit flies d) White flies
20. Lloyd botanical garden is located at:
    a) New Delhi b) Bangalore
    c) Chandigarh d) Darjeeling
21. Which of the following tree species is also known as 'Flame of the Forest'?
    a) *Cassia fistula* b) *Grevillea robusta*
    c) *Butea monospema* d) *Sarca indica*
22. Rock garden is an essential feature of:
    a) Mughal gardens b) Japanese garden
    c) French garden d) English garden
23. In epicotyle grafting of mango, the age of rootstock is:
    a) 1 week b) 2 week
    c) 3 week d) 4 week
24. Sugar concentration in syrup is:
    a) 45ºB b) 55ºB
    c) 65ºB d) 75ºB
25. Persimmon is the national fruit of:
    a) Bangladesh b) Japan
    c) China d) Malaysia
26. Ecosystem in an enclosed glass container is known as:
    a) Aquarium b) Terrarium
    c) Phytotron d) Vivarium
27. Photoblastism is found is:
    a) Tomato b) Cauliflower
    c) Asparagus d) Lettuce
28. Cooling of leafy vegetable is done through:
    a) Air blasting b) Vacuum cooling
    c) Hydro cooling d) Room cooling
29. Which of the following is a drought hardy crop:
    a) Potato b) Tapioca
    c) Asparagus d) Sweet potato

30. Dried stigma is the economic part of:
    a) Clove b) All spice
    c) Kokum d) Saffron
31. 'Pusa Pitamber' a mango hybrid is a cross between:
    a) Amrapali x Sensation b) Dushehari x Sensation
    c) Amrapali x Lal Sundari d) Amrapali x Vanraj
32. The term 'pulsing' is related to:
    a) Fruit preservation b) Cut flower
    c) Pulses d) Dryu flwer
33. 'Golden Rice' is a rich source of:
    a) Vitamin A b) Protein
    c) Lysine d) Oil
34. Which of the following maturity indices is/are used for harvesting of banana?
    a) Disappearance of angularity b) Fullness of cheeks
    c) Specific gravity d) TSS
35. Maturity index of onion used for harvesting is:
    a) Pungency b) Neck fall
    c) Drying of bulbs d) All of these
36. Wind breaks are planted on the side of an orchard:
    a) South-East b) South-West
    c) North-East d) North-West
37. The most suitable plant for biofencing is:
    a) *Capparis decidua* b) *Carrisa congesta*
    c) *Prosopis cineraria* d) *Grewia asiatica*
38. AGMARK is related to:
    a) Marketing of agricultural products
    b) Quality of agricultural products
    c) Processing of agricultural products
    d) Storage of agricultural products
39. Cryopreservation is associated with:
    a) Liquid nitrogen b) Liquid oxygen
    c) Liquid potassium d) Liquid carbon dioxide

40. The most suitable plant tissue for nutrient analysis of pineapple is:
    a) A-leaf b) B-leaf
    c) C-leaf d) D-leaf
41. Which of the following is an important source of calcium?
    a) Lime b) Pyrites
    c) Olivine d) Tourmaline
42. Roses with single large bloom are known as:
    a) Hybrid Tea b) Floribunda
    c) Polyantha d) Miniature
43. Growing of trees for aesthetic purpose is known as:
    a) Silviculture b) Arboriculture
    c) Sericulture d) Horticulture
44. Waxing helps in increasing storage life of fresh fruits by:
    a) Reducing evaporation
    b) Reducing respiration
    c) Reducing evaporation and
    d) Arresting microbial growth respiration
45. Pectin is an essential component of:
    a) Pickle making b) Jam making
    c) Jelly making d) Cordial making
46. Which of the following is not used as a propagating material of pineapple?
    a) Slips b) Suckers
    c) Crown d) Cuttings
47. Which of the following is an ethylene absorbent?
    a) KMno4 b) K2HPO4
    c) KI d) All of these
48. Bt. brinjal has been developed for resistance to:
    a) Little leaf b) Bacterial wilt
    c) Phomopsis blight d) Fruit and shoot borer
49. Triple sigmoid growth curve is found is:
    a) Mango b) Apple
    c) Peach d) Kiwifruit

50. Which of the following acid is found in grapes?
    a) Citric acid b) Malic acid
    c) Tartaric acid d) Ascorbic acid
51. Strawberry is commercially propagated by:
    a) Suckers b) Runners
    c) Stolens d) Cuttings
52. Jamun is recommended for patients suffering from:
    a) Cancer b) Heart disease
    c) Diabetes d) Kidney disorders
53. Grafting in monocots is difficult to perform because of:
    a) Absence of vascular cam bium b) Scattering of vascular bundles
    c) Absense of secondary growth d) All of these
54. Which of the following is known as the "Queen of Bulbous Plants"?
    a) Lilium b) Tuberose
    c) Chrysanthemum d) Gladiolus
55. The first auxin isolated by F. Went in 1928 was:
    a) NAA b) IAA
    c) IBA d) 2,4-D
56. Barbados cherry is a rich source of:
    a) Vitamin A b) Vitamin B
    c) Vitamin C d) Vitamin D
57. Which one of the following metal ions is involved in stomatal movement?
    a) $Zn^{++}$ b) $Mg^{++}$
    c) $Mn^{++}$ d) $K^{+}$
58. Which is not a system of pruning of trees?
    a) Central leader b) Modified leader
    c) Open leader d) Open centre
59. Replication in an experiment means:
    a) The number of blocks
    b) Total number of treatments
    c) The number of times a treatment occurs in an experiment
    d) Degree of freedom

60. Chromosome or chromosome fragment without a centromere is called:
    a) Acentric chromosome
    b) Metacentric chromosome
    c) Telocentric chromosome
    d) Acrocentric chromosome
61. Fruit set in brinjal is observed in:
    a) Medium and true short styled flower
    b) Medium and pseudo short styled flower
    c) Pseudo and true short styled flower
    d) Long and medium style flower
62. The botanical name of peach is:
    a) *Prunus domestica*
    b) *Prunus armeniaca*
    c) *Prunus cerasus*
    d) *Prunus persica*
63. Which of the following is not a symptom of graft incompatibility?
    a) Overgrowth at above or below the graft union
    b) Marked differences in growth rate or vigour of scion and root stock
    c) Graft component breaking apart cleanly at graft union
    d) Dark green foliage with rosseting of leaves
64. Which of the following vegetables is highly tolerant to soil acidity?
    a) Okra
    b) Potato
    c) Cauliflower
    d) Asparagus
65. Marmalade is generally associated with the product made from:
    a) Peach
    b) Apple
    c) Grape
    d) Citrus fruits
66. Which one of the following flowers existed on earth even before evolution of man?
    a) Chrysanthemum
    b) Rose
    c) Carnation
    d) Gladiolus
67. Malling-Merton series of rootstocks are used for propagation of:
    a) Apple
    b) Plum
    c) Peach
    d) Pear
68. 'Cut face' is a physiological disorder of:
    a) Okra
    b) Tomato
    c) Watermelon
    d) Capsicum

69. Which of the following is not a cross-pollinated vegetable?
    a) Radish b) Amaranthus
    c) Pumpkin d) French bean
70. The floribunda roses have been developed for a cross between:
    a) Hybrid Tea x Polyantha b) Hybrid perpetual x Polyantha
    c) Hybrid polyantha x Tea rose d) Hybrid perpetual x Tea rose
71. Which of the following species is not used as rootstock for rose?
    a) *Rose indica* b) *Rosa multiflora*
    c) *Rosa damascena* d) *Rosa bourborniana*
72. In the development of hybrids, availability of male sterile lines has played a significant role in:
    a) Sweet potato b) Onion
    c) Brinjal d) Potato
73. Which of the following is not a bio-pesticide?
    a) Trichogramma b) Spodoptera
    c) Trichoderma d) Bacillus thuringenesis
74. Which of the following vegetables is native to India?
    a) Chilli b) Brinjal
    c) Pumpkin d) Bitter gourd
75. Seed production in cabbage is done by:
    a) Seed to seed method b) Seed plot technique
    c) Head and seed method d) True seed method
76. In coconut, the age of seedling for transplanting should be:
    a) 6-8 months b) 9-12 months
    c) 15-18 months d) 19-24 months
77. Turmeric is multiplied by:
    a) Stem tuber b) Corm
    c) Rhizome d) Tubjrr.us root
78. 'Rajendra Swati is a variety of:
    a) Coriander b) Fenugreek
    c) Black pepper d) Cinnamon

79. Basic chromosome number of gladiolus is:
    a) 15 b) 60
    c) 12 d) 30
80. Pineapple belongs to ...........family:
    a) Annonaceae b) Bromeliaceae
    c) Myrtaceae d) Rosaceae
81. The layout of formal gardens is:
    a) Vertically symmetrical b) Horizontally symmetrical
    c) Bilaterally symmetrical d) Longitudinally symmetrical
82. Bignonia venusta is a flowering climber which blooms in:
    a) Summer b) Rainy
    c) Winter d) Throughout the year
83. Herbaceous boarder is an important feature of:
    a) Formal garden b) Informal garden
    c) Wild garden d) Terrace garden
84. Which of the following fruits is not a multiple fruit?
    a) Jack fruit b) Custard apple
    c) Pineapple d) Mulberry
85. Which of the following chemicals is used for bunch treatment to increase the bunch weight in banana?
    a) NAA b) Urea
    c) KNO3 d) Ammonia
86. The term "Appertizing" is used for:
    a) Dehydration b) Canning
    c) Sterilization d) Syruping
87. The most common method of grape preservation is:
    a) Jam making b) Jelly making
    c) Dehydration d) Wine making
88. Blanching is a treatment given to vegetables before dehydration for:
    a) Destroying enzymes that cause discolouration
    b) For removing air and moisture present in tissues
    c) For improving taste of the product
    d) For maintaining the proper shape of the dehydrated product

89. Orchid seeds are devoid of:
    a) Seed coat b) Cotyledon
    c) Endosperm d) Embryo
90. Alternate bearing is commonly observed in:
    a) Grape b) Orange
    c) Apple d) Pineapple
91. Appearance of male, female and hermaphrodite flowers in the same plant is:
    a) Andromonoecious b) Gynomonoecious
    c) Trimonoecious d) Subandroecious
92. The following organelle of the cell is involved in respiration:
    a) Ribosome b) Mitochondria
    c) Endoplasmic reticulum d) Plasmodesmata
63. 'Bartlett' is a variety of:
    a) Apple b) Peach
    c) Pear d) Plum
94. Growing annual crops in between rows of tree crops is known as:
    a) Cover cropping b) Intercropping
    c) Mixed cropping d) Companion cropping
95. Peach is commercially propagated by:
    a) Grafting b) Air layering
    c) Stooling d) Stem cutting
96. Which of the following is an effect of mulching?
    a) It conserves moisture by reducing evaporation
    b) It keeps the soil warm in winter and cool in summer
    c) It destroys the harmful microbes
    d) It eliminates competition between crop and weeds
97. Scurvy disease in children is caused due to deficiency of:
    a) Vitamin A b) Vitamin B
    c) Vitamin C d) Vitamin D

98. Fleshy underground modified stems that develop horizontally near the soil surface with short nodes and ink-modes are known as:

a) Bulb  b) Rhizome

c) Corm  d) Offset

99. Botanical name of tapioca is:

a) *Ipomoea batatas*  b) *Manihot esculenta*

c) *Xanthosoma sagittifolium*  d) *Cyamopsis tetragonoloba*

100. Which of the following growth regulators is used to induce male flower for maintenance of gynoecious line in cucumber?

a) IAA  b) 2,4-D

c) Ethephone  d) GA3

101. Which of the following vegetables is anemophilous?

a) Cabbage  b) Radish

c) Onion  d) Amaranthus

102. Which is not a feature of garden?

a) Terrace  b) Water basin

c) High wall  d) Wooden gate studded with iron nails

103. Mobility in the garden is created by use of:

a) Trees of different size  b) Evergreen trees

c) Deciduous trees  d) Evergreen shrubs

104. 'Washington Navel' is an important variety of:

a) Sweet orange  b) Mandarin orange

c) Poncirus  d) Fortunella

105. Per-inoculation with mild strain of virus to give protection against infection by virulent strain is known as:

a) Virus indexing  b) Cross protection technique

c) Bud wood certification  d) Rootstock certification

106. Which of the following is not used as loose flower?

a) Rose  b) Chrysanthemum

c) Orchid  d) Tuberose

107. No pinch no stake is a group of plants developed in:

a) Marigold  b) Chrysanthemum

c) Spray carnation  d) Standard carnation

108. Which of the following belongs to the family Rosaceae?
   a) Mango b) Strawberry
   c) Citrus d) Ber

109. Which of the following does not belong to angiosperras?
   a) Mango b) Indian gooseberry
   c) Citrus d) Chilgoza

110. Which of the following fruits even though propagated by seeds, do not have much variability (varietal diversity)?
   a) Litchi b) Mangosteen
   c) Loquat d) Papaya

111. Self-sterility is observed in:
   a) Papaya b) Peach
   c) Apple d) Mango

112. The hormone used to induce flowering in pineapple is:
   a) Auxin b) Cytokinin
   c) Gibberellin d) Ethylene

113. M-27 rootstock of apple is:
   a) Drought resistant b) Ultra dwarf
   c) Woolly apple aphid resistant d) Pumpkin

114. Nectarine is fuzzless mutant of:
   a) Plum b) Almond
   c) Peach d) Apricot

115. Physiological disorder of granulation occurs in:
   a) Mango b) Citrus
   c) Litchi d) Fig

116. Fig is commercially propagated through:
   a) Soft wood cutting b) Semi-soft wood cutting
   c) Hard wood cutting d) Veneer grafting

117. Quince-C rootstock belongs to:
   a) Peach b) Apple
   c) Plum d) Pear

118. Kagzi lime is an indicator plant of:

| a) | Greening | b) | Exocortis |
|---|---|---|---|
| c) | Citrus canker | d) | Tristeza |

119. Summer pruning forms an essential part in the following system of pruning:

| a) | Cordon | b) | Bower system |
|---|---|---|---|
| c) | Dwarf pyramid | d) | Espalier |

120. Dry karonda is the richest source of:

| a) | Calcium | b) | Copper |
|---|---|---|---|
| c) | Iron | d) | Zinc |

121. Litchi flowers are devoid of:

| a) | Stamens | b) | Calyx |
|---|---|---|---|
| c) | Corolla | d) | Pistil |

122. Degreening in citrus is done with the application of:

| a) | GA3 | b) | Ethcphon |
|---|---|---|---|
| c) | Auxins | d) | Cytokinin |

123. The active ingredient of bael having medicinal value is:

| a) | Annonine | b) | Marmelosin |
|---|---|---|---|
| c) | Ftcin | d) | Bromelin |

124. Which of the following mushrooms is edible but not cultivated?

| a) | *Podaxis pistillaris* | b) | *Lentinus edodes* |
|---|---|---|---|
| c) | *Pleurotus sajor-caju* | d) | *Volvariella vohacea* |

125. Which one is a variety of brinjal?

| a) | Pusa Ruby | b) | Pusa Kranti |
|---|---|---|---|
| c) | Pusa Red | d) | Pusa Sadabahar |

126. Physiological disorder 'whiptail' occurs in cauliflower due to the deficiency of:

| a) | Mo | b) | N |
|---|---|---|---|
| c) | Zn | d) | K |

127. Sarpagandha is used to:

| a) | Cure asthma | b) | Cure heart diseases |
|---|---|---|---|
| c) | Treat hypertension | d) | Make general tonic |

128. A product prepared from cabbage by lactic acid fermentation is known as:
   a) Vinegar b) Cider
   c) Pickles d) Sauerkraut
129. The highest vitamin C (mg/100g) is found in following leafy vegetable:
   a) Coriander leaf b) Cabbage
   c) Parsley d) Amaranthus
130. National Bureau of Plant Genetic Resources (NBPGR) is situated at:
   a) Hyderabad b) New Delhi
   c) Lucknow d) Pune
131. Two leaf and a bud is a type of plucking in:
   a) Coffee b) Tea
   c) Palak d) Methi
132. Fruit crop suitable for cold desert of cold arid zone is:
   a) Peach b) Plum
   c) Apple d) Apricot
133. The growth regulator that acts as anti-gibberellins is:
   a) Cytokinin b) Auxin
   c) Morphactin d) Paclobutrazol
134. Causal organism of late blight of potato is:
   a) *Phytopthora infestans* b) *Rhizoctonia solani*
   c) *Fusarium* spp. d) *Erwinia carotovora*
135. Cumin belongs to ................family:
   a) Apiaceae b) Lauraceae
   c) Myrtaceae d) Fabaceae
136. Safed Musli belongs to ........family:
   a) Apiaceae b) Liliaceae
   c) Rubiaceae d) Rosaceae
137. Which part of lemon grass is used for aroma extraction?
   a) Flower b) Roots
   c) Leaves d) Whole plant
138. Which part of Isabgol is used for medicinal purpose?
   a) Husk and Seed b) Root and Leaves
   c) Bark d) Roots

139. Anthurium is generally propagated through:

a) Runners b) Suckers

c) Cutting d) Bulbs

140. FPO/FSSAI specifications for fruit jelly and marmalade are:

a) TSS-65%, Juice-45% b) TSS-68%, Juice-55%

c) TSS-68%, Juice-45% d) TSS-50%, Juice-40%

141. Which of the following techniques is used for detection of viruses?

a) FL1SA b) AFLP

c) RFLP d) RAPD

142. The first Agricultural University of India is located at:

a) Coimbatore b) Pantnagar

c) Samastipur d) Punjab

143. The loss in vigor and productivity of plant due to selfing is known as:

a) Inbreeding depression b) Heterosts

c) Clonal degeneration d) Luxuriance

144. Indian Institute of Vegetable Research is situated at:

a) Kanpur b) Nagpur

c) Varanasi d) Chennai

145. Floating garden of Dal lake of Kashmir was developed at the base of:

a) Typha grass b) Water lilies

c) Mud and sand d) Dal weeds

146. Seed plot technique is an ideal method used in:

a) Seed production of cauliflower

b) Seed production of peas and beans

c) Seed production of cucurbits

d) Production of seed potatoes

147. The national pickle of India is:

a) Mango b) Lime

c) Aonla d) Cucumber

148. Which of the following is a perennial cucurbits vegetable:

a) Pumpkin b) Ridge gourd

c) Ash gourd d) Chayote

149. Which of the following seed rate of TPS is recommended for planting one ha area?

a) 150g b) 250g
c) 350g d) 50g

150. Rutabaga is a cross of:

a) *Brassica nigra* x *B. okracea*
b) *Brassica nigra* x *B. juncea*
c) *Brassica okracea* x *B. compestris*
d) None of these

151. Match the following:

| | | | |
|---|---|---|---|
| i) | CIMAP | a) | New Delhi |
| ii) | IIHR | b) | Bangalore |
| iii) | CSIR | c) | Mysore |
| iv) | CTCRI | d) | Lucknow |
| v) | CFTRI | e) | Thiruvananthapuram |

152. Match the following:

| | | | |
|---|---|---|---|
| i) | Cell division | a) | Cytokinin |
| ii) | Overcome dwarfism | b) | Gibberellins |
| iii) | Apical dominance | c) | Auxins |
| iv) | Fruit ripening | d) | Ethylene |
| v) | Closing of stomata | e) | Abscisic acid |

153. Match the following:

| | | | |
|---|---|---|---|
| i) | Scarification | a) | Guava |
| ii) | Stratification | b) | Temperate fruits |
| iii) | Overcome dormancy | c) | Gibbetefiias |
| iv) | Thiourea | d) | Potato |
| v) | Hot water treatment | e) | Cole crops |

154. Match the following:

| | | | |
|---|---|---|---|
| i) | Square system | a) | Terrace system |
| ii) | Diagonal system | b) | Practiced in hilly areas |
| iii) | Hexagonal system | c) | Practiced near cities |
| iv) | Contour system | d) | Filler plant |
| v) | Check soil erosion | e) | Simplest |

155. Match the following:

| | | | |
|---|---|---|---|
| i) | Annual | a) | Kochia |
| ii) | Shrub | b) | Karonda |
| iii) | Climber | c) | Railway creeper |
| iv) | Tree | d) | Tuberose |
| v) | Bulbous plant | e) | Palash |

156. Match the following:

| | | | |
|---|---|---|---|
| i) | Malformation | a) | Strawberry |
| ii) | Shot berry | b) | Papaya |
| iii) | Choke throat | c) | Banana |
| iv) | Skin freckling | d) | Sapota |
| v) | Cocks comb | e) | Grape |

157. Match the following:

| | | | |
|---|---|---|---|
| i) | English garden | a) | Buddha jayanti garden |
| ii) | Japanese | b) | Rockery |
| iii) | Mughal garden | c) | Tajmahal |
| iv) | Garden adornment | d) | Seat |
| v) | Garden style | e) | Informal |

158. Match the following:

| | | | |
|---|---|---|---|
| i) | Apple | a) | Triple sigmoidal growth curve |
| ii) | Fig | b) | Cutting |
| iii) | Kiwifruit | c) | Single sigmoidal growth curve |
| iv) | Pomegranate | d) | Thorn less species |
| v) | Pummelo | e) | Double sigmoidal growth curve |

159. Match the following:

| | | | |
|---|---|---|---|
| i) | Black heart | a) | Cucumber |
| ii) | Golden fleck | b) | Radish |
| iii) | Pillow | c) | Cauliflower |
| iv) | Whiptail | d) | Potato |
| v) | Pithiness | e) | Tomato |

160. Match the following:

| | | | |
|---|---|---|---|
| i) | Star apple | a) | Tiliaceae |
| ii) | Star fruit | b) | Oxalidaceac |
| iii) | Mangosteen | c) | Rhamnaceae |
| iv) | Loquat | d) | Clusiaceae |
| v) | Ber | e) | Rosaceae |

## Answers Key

| | | | | | | | | | | | | | |
|---|---|---|---|---|---|---|---|---|---|---|---|---|---|
| 1. | (b) | 2. | (a) | 3. | (b) | 4. | (d) | 5. | (a) | 6. | (b) | 7. | (a) |
| 8. | (a) | 9. | (c) | 10. | (a) | 11. | (d) | 12. | (c) | 13. | (b) | 14. | (d) |
| 15. | (b) | 16. | (b) | 17. | (a) | 18. | (b) | 19. | (d) | 20. | (d) | 21. | (c) |
| 22. | (b) | 23. | (b) | 24. | (c) | 25. | (b) | 26. | (b) | 27. | (d) | 28. | (b) |
| 29. | (b) | 30. | (d) | 31. | (c) | 32. | (b) | 33. | (a) | 34. | (a) | 35. | (b) |
| 36. | (d) | 37. | (b) | 38. | (b) | 39. | (a) | 40. | (d) | 41. | (a) | 42. | (a) |
| 43. | (b) | 44. | (b) | 45. | (c) | 46. | (d) | 47. | (a) | 48. | (d) | 49. | (d) |
| 50. | (c) | 51. | (b) | 52. | (c) | 53. | (a) | 54. | (d) | 55. | (b) | 56. | (c) |
| 57. | (d) | 58. | (c) | 59. | (c) | 60. | (a) | 61. | (d) | 62. | (d) | 63. | (d) |
| 64. | (b) | 65. | (d) | 66. | (b) | 67. | (a) | 68. | (b) | 69. | (d) | 70. | (a) |
| 71. | (c) | 72. | (b) | 73. | (b) | 74. | (b) | 75. | (c) | 76. | (b) | 77. | (c) |
| 78. | (a) | 79. | (a) | 80. | (b) | 81. | (c) | 82. | (c) | 83. | (c) | 84. | (b) |
| 85. | (c) | 86. | (b) | 87. | (c) | 88. | (a) | 89. | (c) | 90. | (c) | 91. | (c) |
| 92. | (b) | 93. | (c) | 94. | (b) | 95. | (a) | 96. | (c) | 97. | (c) | 98. | (b) |
| 99. | (b) | 100. | (d) | 101. | (d) | 102. | (d) | 103. | (c) | 104. | (a) | 105. | (b) |
| 106. | (c) | 107. | (b) | 108. | (b) | 109. | (d) | 110. | (b) | 111. | (c) | 112. | (d) |
| 113. | (b) | 114. | (c) | 115. | (b) | 116. | (c) | 117. | (d) | 118. | (d) | 119. | (d) |
| 120. | (c) | 121. | (c) | 122. | (b) | 123. | (b) | 124. | (a) | 125. | (b) | 126. | (a) |
| 127. | (a) | 128. | (d) | 129. | (a) | 130. | (b) | 131. | (b) | 132. | (d) | 133. | (d) |
| 134. | (a) | 135. | (a) | 136. | (b) | 137. | (c) | 138. | (a) | 139. | (b) | 140. | (a) |
| 141. | (a) | 142. | (b) | 143. | (a) | 144. | (c) | 145. | (a) | 146. | (d) | 147. | (b) |
| 148. | (d) | 149. | (a) | 150. | (c) | | | | | | | | |

| Q. | 151. | 152. | 153. | 154. | 155. | 156. | 157. | 158. | 159. | 160. |
|---|---|---|---|---|---|---|---|---|---|---|
| i) | (d) | (a) | (a) | (e) | (a) | (a) | (b) | (c) | (d) | (a) |
| ii) | (b) | (b) | (b) | (d) | (b) | (e) | (a) | (e) | (e) | (b) |
| iii) | (a) | (c) | (c) | (c) | (c) | (c) | (c) | (a) | (a) | (d) |
| iv) | (e) | (d) | (d) | (b) | (e) | (b) | (d) | (b) | (c) | (e) |
| v) | (c) | (e) | (e) | (a) | (d) | (d) | (e) | (d) | (b) | (c) |

# 6

# ICAR – JRF Horticulture Exam – 2015

1. The end product of glycolysis is:
   a) Malic acid  b) Citric acid
   c) Pyruvic acid  d) Oxalic acid
2. Tag colour of certified seed is:
   a) White  b) Purple
   c) Blue  d) Golden yellow
3. Which of the following varietyiesof sweet potato is rich in carotene content?
   a) Gauri  b) Rajendra Shakarkand-35
   c) Pusa Sundri  d) Rajendra Shakarkand-43
4. India is the largest producer of:
   a) Mango  b) Banana
   c) Papaya  d) All of these
5. CIAH is located at:
   a) Jodhpur  b) Bikaner
   c) Lucknow  d) Anand
6. CIMAP is situated at:
   a) Lucknow  b) Bangalore
   c) Hyderabad  d) Anand
7. Sex form of Pusa Delicious variety of papaya is:
   a) Dioecious  b) Gynodioecious
   c) Monoecious  d) Androdioecious
8. Which of the following type sof flowers is/are present in coconut?
   a) Monoecious  b) Dioecious
   c) Hermaphrodite  d) All of these

9. Which of the following fruits is used to cure high blood pressure?
   a) Jamun  b) Bael
   c) Avocado  d) Apple
10. High water table is a limiting factor for cultivation of:
   a) Banana  b) Papaya
   c) Pineapple  d) Guava
11. Which of the following is a seedless variety of mango:
   a) Amarapali  b) Arka Aruna
   c) Sindhu  d) Kesar
12. Bordeaux mixture was first discovered by:
   a) P. M. A. Millardet  b) Kogl *et al.*
   c) G. J. Mendel  d) Zimmerman and Wilcoxon
13. Naveen, Vaishali, Rupali, Avinash and Krishna are the varieties of:
   a) Brinjal  b) Tomato
   c) Chilli  d) Okra
14. TSS of jam should be:
   a) 70%  b) 45%
   c) 35%  d) 68%
15. Sugar content in finished jelly should be:
   a) 70%  b) 45%
   c) 35%  d) 65%
16. Phalsa belongs to the family of:
   a) Rosaceae  b) Tilliaceae
   c) Myrtaceae  d) Guttiferae
17. The pressure maintained in high pressure drip irrigation system is:
   a) 30 psi  b) <30 psi
   c) >30 psi  d) None of these
18. Bending of shoots is practiced in:
   a) Dahlia  b) Carnation
   c) Chrysanthemum  d) Roses in greenhouse
19. Terminal cutting of chrysanthemum is done during:
   a) End of June  b) July - August
   c) September - October  d) November - December

20. Which of the following varieties of aonla is free from fruit necrosis?
    a) Francis b) Chakaiya
    c) Banarasi d) MA-9
21. Plant Quarantine Act came into force in:
    a) 1914 b) 1941
    c) 1966 d) 1973
22. Birbal Sahni is a variety of:
    a) Carnation b) Chrysanthemum
    c) Rose d) Dahlia
23. Pusa Narangi is a variety of:
    a) Rose b) Marigold
    c) Jasmine d) Tuberose
24. Which one of the following is sensitive to ethylene injury?
    a) Gerbera b) Tulip
    c) Anthurium d) Antirrhinum
25. Harvesting of standard carnation is done at:
    a) Full open b) Half open
    c) Tight bud stage d) Painting brush stage
26. Isolation distance required for cross pollinated ornamentals is:
    a) 400 m b) 500 m
    c) 600 m d) 1000 m
27. Pear is trained through:
    a) Central leader system b) Open centre system
    c) Modified central leader system d) Trellis system
28. Which one of the following is a winter hardy rootstock of apple?
    a) M-16 b) MM-104
    c) M-27 d) M-4
29. Which disease occurs in HDP of banana?
    a) Sigatoka leaf spot b) Finger tip
    c) Bunchy top d) Panama wilt
30. Which of the following compounds is found in radish?
    a) Non adienal b) Dimethyl disulphide
    c) Dimethyl pyarzin d) Isothiocynate

31. Which type of polyembryony is present in mango?
   a) Vegetative  b) Parthenogenesis
   c) Non-recurrent  d) Nucellar polyembryony
32. Soil condition favourable for growth of nematodes is:
   a) Heavy soils  b) Soil rich in organic matter
   c) Sandy loam soils  d) No reaction with soil
33. Fruit plants are generally irrigated ...........stage of growth:
   a) Flowering to fruiting  b) Fully mature stage
   c) Fully ripe stage  d) At the time of flower emergence
34. Which one the following is not a cane pruned variety of grape?
   a) Gulabi  b) Pusa Seedless
   c) Thompson Seedless  d) Bangalore Blue
35. Which of the following is a non-climacteric fruit?
   a) Apple  b) Annona
   c) Peach  d) Pineapple
36. Which of the following is a physiological disorder of tomato?
   a) Puffmess  b) Hollow stem
   c) Black heart  d) Splitting
37. 'Stone fruit' formation is a physiological disorder of:
   a) Annona  b) Pineapple
   c) Peach  d) Plum
38. Jelly seed is a physiological disorder of:
   a) Grape  b) Mango
   c) Sapota  d) Guava
39. Mahrun, Katha, Dodhia, Jogia, Kadaka are the varieties of:
   a) Ber  b) Bael
   c) Aonla  d) Almond
40. Which of the following is a dicot fruit crop?
   a) Banana  b) Pineapple
   c) Date palm  d) Guava
41. For tuberization of potato, the night temperature should be:
   a) Below 23°C  b) Below 17°C
   c) Below 30°C  d) Above 30°C

42. Constituent of potato is not suitable for processing:
    a) High dry matter b) High phenol and sugar
    c) Low phenol d) Low sugar
43. Fruit drop in citrus can be controlled by application of:
    a) Auxin b) Cytokinin
    c) Gibberellins d) ABA
44. Which of the following elements plays an important role in auxin biosynthesis?
    a) Mn b) S
    c) Zn d) Cu
45. Metaxenia is normally found in which of the following fruit crops:
    a) Aonla b) Ber
    c) Date palm d) Fig
46. Sickle shaped leaves in gladiolus is a symptom of:
    a) Stem rot b) Bacterial blight
    c) Root rot d) Collar rot
47. The economic part of clove is:
    a) Flower b) Unopened flower bud
    c) Fruit d) Stigma
48. *Mimusops hexendra* is a rootstock of:
    a) Sapota b) Loquat
    c) Apple d) Peach
49. Which of the following varieties of guava is tolerant to wilt disease?
    a) Allahabad Safeda b) Sardar
    c) Allahabad Round d) Hafsi
50. Active principle compound present in black pepper is:
    a) 1,8 cineol b) Piperine
    c) Anethol d) Phenol
51. Photosynthetically active radiation (PAR) wave length range is:
    a) 100 to 300 nm b) 400 to 700 nm
    c) 700 to 1000 nm d) 1000 to 13000 nm

52. Nectarines are smooth skinned (fuzzless) mutants of:
    a) Pear  b) Plum
    c) Peach  d) Pineapple
53. Coated or pelleted seeds are used for:
    a) Low germination seed rate  b) Bold seeds
    c) Prevention of odour  d) Dormant seeds
54. Which of the following weeds leads to asthma?
    a) Phalaris minor  b) Argimone mexicana
    c) Franseria acanthicarpa  d) Parthenium hysterophorus
55. The alcohol content in natural wine varies from:
    a) 6-16%  b) 10-20%
    c) 15-25%  d) 20-30%
56. "Flat sour" spoilage in canned food is caused by:
    a) Bacillus steriothermophilus
    b) Clofstridium nigrificans
    c) Clostridium thermosacharolyticum
    d) E. coli
57. Blossom end rot in grape can be associated with the deficiency of which of the following:
    a) Zinc  b) Boron
    c) Sulphur  d) Calcium
58. Greening disease of citrus is transmitted by:
    a) Citrus whitefly  b) Cottony cushion scale
    c) Citrus psylla  d) Citrus leaf minor
59. The most pungent part of chilli is:
    a) Placenta  b) Seeds
    c) Pericarp  d) Peduncle
60. Who discovered student's t-test?
    a) S. D. Poisson  b) R. A. Fisher
    c) W. S. Cosset  d) Karl Pearson
61. Defective broken bits of coffee seeds are known as:
    a) Waste  b) Black
    c) Triage  d) Splits

62. The type of storage for equalization of moisture to a desired uniform level in dehydrated fruits is

| | | | |
|---|---|---|---|
| a) | Holding | b) | Blowing |
| c) | Smudging | d) | Sweating |

63. For processing purpose, potatoes are best stored at:

| | | | |
|---|---|---|---|
| a) | 3°C | b) | 7°C |
| c) | 10°C | d) | 14°C |

64. The minimum specific gravity of mango fruits at the time of harvesting should be:

| | | | |
|---|---|---|---|
| a) | 0.98 | b) | 1.00 |
| c) | 1.02 | d) | 1.04 |

65. Litchi is commercially propagated by:

| | | | |
|---|---|---|---|
| a) | Suckers | b) | Ground layering |
| c) | Cuttings | d) | Air layering |

66. Sabudana/Sago is a product prepared from:

| | | | |
|---|---|---|---|
| a) | Arrowroot | b) | Potato |
| c) | Cassava | d) | Coleus |

67. Kinnow mandarin is a cross between:

| | | | |
|---|---|---|---|
| a) | C. nob His x C. deliciosa | b) | C. sinensis x C. nobilis |
| c) | C. nobilis x C. sinensis | d) | C. deliciosa x C. nobilis |

68. With respect to mode of action of ethylene in plants, the ethylene binds to an active site most probably containing:

| | | | |
|---|---|---|---|
| a) | Silver | b) | Copper |
| c) | Iron | d) | Manganese |

69. Which is a post-harvest physiological disorder of mango:

| | | | |
|---|---|---|---|
| a) | Black tip | b) | Internal necrosis |
| c) | Stem end cavity | d) | Sap burn |

70. Crossing over takes place in which of the following stages of meiosis,:

| | | | |
|---|---|---|---|
| a) | Pachytene | b) | Zygotene |
| c) | Leptotene | d) | Diplotene |

71. Breeding method used to transfer disease resistance:

| | | | |
|---|---|---|---|
| a) | Pedigree selection | b) | Bulk method |
| c) | Pureline selection | d) | Back cross method |

72. The latest material used in refrigerated storages for insulation is:
    a) Thermo coal b) Cork board
    c) Polystyrene d) Poly urethane foam
73. Cashewnuts has protein content (%) of:
    a) 11 b) 16
    c) 21 d) 26
74. Which of the following is a seed spice?
    a) Black pepper b) Clove
    c) Ginger d) Cumin
75. The precursor of ethylene biosynthesis in plants is:
    a) Cysteine b) Valine
    c) Methionine d) Proline
76. 'Screwpine' is the common name of which aromatic plant:
    a) Mentha b) Kewda
    c) Citronella d) Lemon grass
77. Fruits of Pusa Naveen variety of bottle gourd are:
    a) Perfectly round b) Long with crooked neck
    c) Cylindrical without crook neck d) Long with slight curve
78. The predominant acid found in tamarind is:
    a) Citric acid b) Malic acid
    c) Acetic acid d) Tartaric acid
79. Which of the following vegetable crops has a shallow root system?
    a) Tomato b) Pea
    c) Carrot d) Lettuce
80. In tomato, the vector of leaf curl virus is:
    a) Aphids b) Leaf hoppers
    c) Whitefly d) Thrips
81. 'Ratna' variety of mango is a cross between:
    a) Alphonso x Banganpalli b) Nee/um x Langra
    c) Neelum x Alphonso d) Dashehari x Neelum
82. The cause of blossom end rot tomato is:
    a) Fungus b) Virus
    c) Bacteria d) Physiological disorder

83. 'Fruits, flowers and foliage are used in which of the following arrangement:

a) Ikenobo b) Morimono

c) Nageire d) Zenei-ka

84. Starch test is performed for the harvest maturity of:

a) Apple b) Guava

c) Banana d) Custard apple

85. Blanching is a pre-treatment which is mainly done to:

a) Soften the produce b) Kill all the micro-organ ism

c) Increase the sweetness d) Inactivate the enzymes

86. Binomial nomenclature was suggested by:

a) Linnaeus b) Griffith

c) Mendel d) Hutchinson

87. Strawberry fruits are:

a) Aggregate b) Multiplt

c) Simple d) Berriei

88. The vegetable which is an important source of iodine inhuman diet?

a) Tapioca b) Okra

c) Tomato d) Spinach

89. Which is not essentially a feature of English garden?

a) Lawn b) Stone lanterns

c) Herbaceous border d) Rockery

90. The shrub used for ornamostal foliage is:

a) Thuja b) Hibiscus

c) Ixora d) Hamelia patens

91. Hollow heart is a physiological disorder is associated with:

a) Onion b) Turnip

c) Potato d) Carrot

92. Ultra dwarfing root stock of apple is:

a) M-9 b) M-13

c) Baldwin d) M-27

93. Shoot and fruit borer is an important pest of:

a) Tomato b) Chilli

c) Potato d) Brinjal

94. Best planting material in pineapple is:
   a) Crown b) Slips
   c) Suckers d) Seed
95. How many extra plants can be accommodated in hexagonal system of layout?
   a) 15% b) 25%
   c) 10% d) 30%
96. Richest source of vitamin C is:
   a) Aonla b) Citrus
   c) Barbados cherry d) Karonda
97. Following is the indicator plant for boron and molybdenum deficiency:
   a) Cauliflower b) Potato
   c) Brinjal d) Chilli
98. Following component is involved in ATP and ADP energy transformation in plants:
   a) Calcium b) Phosphorus
   c) Potassium d) Magnesium
99. The quantity of brine required for canning of vegetables is:
   a) 1 to 3% b) 3 to 6%
   c) 6 to 9% d) 9 to 12%
100. ................ is a seedless watermelon variety is:
   a) New Hampshire Midgut b) Arka Madhura
   c) Tetrad-2 d) Tetrad-4
101. ............... *is an in-situ* method of mango grafting:
   a) Stooling b) Soft wood grafting
   c) Approach grafting d) Inarching
102. Enzymatic discoloration in potatoes can be prevented by the use of:
   a) Nitrogenous fertilizers b) DAP
   c) Murate of potash d) SSP
103. Pomegranate fruit is botanically called 'Balusta'. This is a modified form of:
   a) Drupe c) Berry
   b) Stone d) Nut

104. Which of the following flowers is highly sensitive to chilling injury?

a) Tulip b) Lily

c) Anthurium d) Freesia

105. Which of the following is a typical example of CAM plant:

a) Coconut b) Banana

c) Pineapple d) Arecanut

106. The acidity per cent in terms of citric acid in a good quality tomato juice should be:

a) 0.2 b) 0.4

c) 0.6 d) 0.8

107. Process of removal of air from a filled can is termed as:

a) Blanching b) Exhausting

c) Processing d) Sealing

108. Expanded form of PCR is:

a) Polymerase Chain Reaction b) Polymorphic Character Regulator

c) Polymorphic Catalytic Regulator d) Polymorphic Chain Reaction

109. Lincoln is an exotic variety of:

a) Cowpea b) Pea

c) Dolichos bean d) French bean

110. The maximum temperature of the dehydrator for the dehydration of vegetables is:

a) 56°C b) 66°C

c) 76°C d) 86°C

111. 'The Queen of the Night' is the common name for:

a) Cestrum diurnam b) Cestrum nocturnum

c) Cestrum elegans d) Cestrum pupurewn

112. Sweet potato belongs to .............family:

a) Euphorbiaceae b) Leguminoceae

c) Convolvulaceae d) Cucurbitaceae

113. The residual moisture in dehydrated vegetables should not be more than:

a) 2 to 4% b) 4 to 6%

c) 6 to 8% d) 8 to 10%

114. Cross protection technique in citrus is effective to check:

a) Mycoplasma b) Fusarium oxysporum

c) Phytopthora palmivora d) Tritesza

115. Quickest method to develop a lawn is:

a) Seeding b) Turfing

c) Turf plastering d) Dibbing

116. A semi-solid transparent product prepared from clear extract of pectin is known as:

a) Jam b) Jelly

c) Marmalade d) Squash

117. Coconut attains full maturity from fruit set within:

a) 6-8 months b) >16 months

c) 10-12 months d) 13-15 months

118. Which cut flower is most sensitive to ethylene?

a) Anthurium b) Carnation

c) Asparagus d) Gerbera

119. The country of origin of Cashew nut is:

a) Brazil b) Caspian sea area

c) India d) South East Asia

120. In tomato, boron deficiency leads to:

a) Puffiness b) Cracking

c) Blossom end rot d) Cat face

121. Which cut flower has the largest cultivation area in India?

a) Rose b) Gladiolus

c) Chrysanthemum d) Orchids

122. An ornamental tree having blue flowers is:

a) Cassia nodosa b) Cassia biflora

c) Jacaranda mimosifolia d) Butea monosperma

123. A carrot variety rich in carotene developed by IARI, Katrain is:

a) Pusa Meghali b) Pusa Yamadagni

c) Top Weight d) Pusa Nishant

124. Which of the following glucosides is present in bitter gourd?

a) Cucurbitacin b) Quercetin
c) Sinigrin d) Allin

125. The chief active ingredient of the medicinal value of the bael is:

a) Annonine b) Marmelosin
c) Ficin d) None of these

126. Peeling of fruits/vegetables using NaOH is termed as:

a) Flame peeling b) Shelling
c) Lye peeling d) Scrapping

127. The process of giving a permanent framework to plants is referred to as:

a) Training b) Staking
c) Pruning d) Nipping

128. Delayed harvesting in radish may cause:

a) Forking b) Pithiness
c) Deformed root d) Misshaping

129. Which of the following is a good pollinizer for apple?

a) Red Delicious b) Royal Delicious
c) Golden Delicious d) Jonathan

130. The first website for auction of flowers in India is in:

a) Delhi b) Pune
c) Bengaluru d) Hyderabad

131. An essential component of proteins, protoplasm and chlorophyll is:

a) Nitrogen b) Potassium
c) Phosphorus d) Sulphur

132. Which of the following does not belong to cole crops?

a) Cabbage b) Cauliflower
c) Kale d) Turnip

133. The Per capita consumption of cut flowers is highest in:

a) India b) Germany
c) USA d) Swil/crland

134. Which of the following crop docs not belong to the family yjngiberaccae?

a) Turmeric b) Ginger
c) Cardamom d) Garlic

135. Type of self-incompatibility in pineapple is:
   a) Gamctophytic b) Sporophytic
   c) Mctcromorphic d) None of these
136. The method of food preservation at 0-5°C is known as:
   a) Refrigeration b) Pasteurization
   c) Freezing d) Sterilization
137. The chemical used to induce male-ness in gynoccious line of cucumber:
   a) NAA b) ABA
   c) Mn d) Silver nitrate
138. Which type of fig bears Breba crop?
   a) Sanpedro fig b) Common fig
   c) Smyrna fig d) Capri fig
139. Which of these is not a type of Japanese garden?
   a) Mill garden b) Terrace garden
   c) Flat garden d) Tea garden
140. The hottest chilli is:
   a) Byadagi Kaddi b) I'usa Jwala
   c) Blurt Jolokia d) Byadagi Dabba
141. Chemical defoliation in roses is done by:
   a) Urea b) Copper sulphate
   c) Auxin d) GA3
142. Type of inflorescence in litchi is:
   a) Raceme b) Umbel
   c) Panicle d) Cymose
143. Which part of Isabgol is used for medicinal purpose?
   a) Husk and Seed b) Root and Leaves
   c) Bark d) Roots
144. Which of the following causes pungency in radish?
   a) Diallyl disulphide b) Isothiocynatc
   c) Allyl propyl disulphide d) Erucic acid
145. Arka Pitamber is a variety:
   a) Yellow colored onion b) White colored onion
   c) Red colored onion d) Kharif onion

146. Seed rate of cauliflower is :

a) 300/400g/ha. b) 400-700g/ha

c) 700-1000g/ha d) 1000-1200g/ha

147. Rio- De- Janerlo it a variety of:

a) Black pepper b) Fenugreek

c) Ginger d) Turmeric

148. How much isolation distance is maintained for foundation seed production in onion?

a) 400 m b) 500 m

c) 800 m d) 1000 m

149. Safed Musali belongs to ..........genus:

a) Chlorophylum b) Ancthum

c) Cassia d) Cephaelis

150. Pasteurization of fruits is done at:

a) 100°C for 30 minutes b) 100°C for 90 minutes

c) 116°C for 30 minutes d) 116°C for 90 minutes

151. Match the following:

i) Okra a) Malvaceae

ii) Tomato b) Chenopodia- -ceae

iii) Spinach c) Compositeae

iv) Lettuce d) Liliaceac

v) Asparagus e) Solanaceae

152. Match the following:

i) Vitamin A a) Scurvy

ii) Vitamin D b) Present in bones and teeth

iii) Vitamin K c) Anti- hcmorrhagie

iv) Vitamin C d) Rickets

v) Ca e) Night blindness

153. Match the following:

i) Pinjore garden a) Kodaikanal

ii) Mandore garden b) New Delhi

iii) Buddha Jayanti Park c) Mysore

iv) Brindavan Garden d) Jodhpur

v) Byrant Park e) Chandigarh

154. Match the following:

| | | | |
|---|---|---|---|
| i) | King of spices | a) | *Oryctes rhinoceros* |
| ii) | Queen of spices | b) | *Anacardium occidentale* |
| iii) | Coconut | c) | Black pepper |
| iv) | Cashew nut | d) | Coconut |
| v) | Monoecious | e) | Small cardamom |

155. Match the following:

| | | | |
|---|---|---|---|
| i) | Mango | a) | Softnose |
| ii) | Tomato | b) | Cat face |
| iii) | Cauliflower | c) | Dry neck |
| iv) | Potato | d) | Buttoning |
| v) | Avocado | e) | Black heart |

156. Match the following:

| | | | |
|---|---|---|---|
| i) | Sugar beet | a) | Saponine |
| ii) | Papaya | b) | Custard apple |
| iii) | Mangold | c) | Luttein |
| iv) | Spinach | d) | Caricaxauthin |
| v) | Aimoiuno | e) | Beta-cyanine |

157. Match the following:

| | | | |
|---|---|---|---|
| i) | Aswagandha | a) | Liliaceae |
| ii) | Sarpagandha | b) | Labiateae |
| iii) | Liquorice | c) | Leguminoseae |
| iv) | Patchouli | d) | Apocynaceae |
| v) | Aloe | e) | Solanaceae |

158. Match the following:

| | | | |
|---|---|---|---|
| i) | Sugar loaf | a) | Mango |
| ii) | Akshay | b) | Pear |
| iii) | Arka Mridtilu | c) | Guava |
| iv) | Punjab Gold | d) | Lemon |
| v) | Eureka | e) | Pineapple |

159. Match the following:

| | | | |
|---|---|---|---|
| i) | Grape | a) | Ellagic acid |
| ii) | Orange | b) | Tartaric acid |
| iii) | Carambola | c) | Citric acid |
| iv) | Strawberry | d) | Malic acid |
| v) | Apple | e) | Oxalic acid |

160. Match the following:

| | | | |
|---|---|---|---|
| i) | Apple | a) | Syconus |
| ii) | Peach | b) | Berry |
| iii) | Litchi | c) | Drupe |
| iv) | Grape | d) | Nut |
| v) | Fig | e) | Pome |

## Answers Key

| | | | | | | | | | | | | | |
|---|---|---|---|---|---|---|---|---|---|---|---|---|---|
| 1. | (c) | 2. | (c) | 3. | (c) | 4. | (d) | 5. | (b) | 6. | (a) | 7. | (b) |
| 8. | (a) | 9. | (a) | 10. | (d) | 11. | (c) | 12. | (a) | 13. | (b) | 14. | (d) |
| 15. | (d) | 16. | (b) | 17. | (a) | 18. | (d) | 19. | (a) | 20. | (b) | 21. | (a) |
| 22. | (b) | 23. | (b) | 24. | (d) | 25. | (d) | 26. | (b) | 27. | (c) | 28. | (b) |
| 29. | (b) | 30. | (d) | 31. | (d) | 32. | (c) | 33. | (a) | 34. | (d) | 35. | (d) |
| 36. | (a) | 37. | (a) | 38. | (b) | 39. | (a) | 40. | (d) | 41. | (a) | 42. | (b) |
| 43. | (a) | 44. | (c) | 45. | (c) | 46. | (d) | 47. | (b) | 48. | (a) | 49. | (b) |
| 50. | (d) | 51. | (b) | 52. | (c) | 53. | (a) | 54. | (c) | 55. | (a) | 56. | (a) |
| 57. | (d) | 58. | (c) | 59. | (a) | 60. | (c) | 61. | (c) | 62. | (d) | 63. | (c) |
| 64. | (c) | 65. | (d) | 66. | (c) | 67. | (a) | 68. | (b) | 69. | (d) | 70. | (a) |
| 71. | (d) | 72. | (a) | 73. | (c) | 74. | (d) | 75. | (c) | 76. | (b) | 77. | (c) |
| 78. | (d) | 79. | (d) | 80. | (c) | 81. | (c) | 82. | (d) | 83. | (b) | 84. | (a) |
| 85. | (d) | 86. | (a) | 87. | (a) | 88. | (b) | 89. | (b) | 90. | (a) | 91. | (c) |
| 92. | (d) | 93. | (d) | 94. | (b) | 95. | (a) | 96. | (c) | 97. | (a) | 98. | (b) |
| 99. | (a) | 100. | (b) | 101. | (b) | 102. | (c) | 103. | (c) | 104. | (c) | 105. | (c) |
| 106. | (b) | 107. | (b) | 108. | (a) | 109. | (b) | 110. | (b) | 111. | (b) | 112. | (c) |
| 113. | (c) | 114. | (d) | 115. | (b) | 116. | (b) | 117. | (c) | 118. | (b) | 119. | (a) |
| 120. | (b) | 121. | (c) | 122. | (c) | 123. | (b) | 124. | (a) | 125. | (b) | 126. | (c) |
| 127. | (a) | 128. | (b) | 129. | (c) | 130. | (c) | 131. | (a) | 132. | (d) | 133. | (d) |
| 134. | (d) | 135. | (a) | 136. | (a) | 137. | (d) | 138. | (a) | 139. | (b) | 140. | (c) |
| 141. | (b) | 142. | (c) | 143. | (a) | 144. | (b) | 145. | (a) | 146. | (b) | 147. | (c) |
| 148. | (d) | 149. | (a) | 150. | (a) | | | | | | | | |

| Q. | 151. | 152. | 153. | 154. | 155. | 156. | 157. | 158. | 159. | 160. |
|---|---|---|---|---|---|---|---|---|---|---|
| i) | (a) | (e) | (e) | (c) | (a) | (e) | (e) | (e) | (b) | (e) |
| ii) | (e) | (d) | (d) | (e) | (b) | (d) | (d) | (a) | (c) | (c) |
| iii) | (b) | (c) | (b) | (a) | (d) | (c) | (c) | (c) | (e) | (d) |
| iv) | (c) | (a) | (c) | (b) | (e) | (a) | (b) | (b) | (a) | (b) |
| v) | (d) | (b) | (a) | (d) | (c) | (b) | (a) | (d) | (d) | (a) |

# 7

# ICAR – JRF Horticulture Exam – 2016

1. Which of the following is a variety of black carrot?
   a) Pusa Meghali  b) Pusa Asita
   c) Pusa Rudhira  d) Pusa Vasuda
2. NRC for orchid is situated at:
   a) Pakyong  b) Gangtok
   c) Lachen  d) Aritar
3. Bonneville variety of pea has been introduced from:
   a) USA  b) Germany
   c) Sweden  d) UK
4. Which of the following fruit has the lighest (mg/100g) vitamin C content?
   a) Barbados cherry  b) Aonla
   c) Guava  d) Citrus
5. The term 'Climacteric' was first used by:
   a) Kidd and West  b) Gane
   c) Cruess  d) Bleekar
6. 'For harvesting Delicious apple at light maturity, the Starch Pattern Index (SPI) should be:
   a) 2.5/10  b) 3.5/10
   c) 4.5/10  d) 6.5/10
7. Which of the following is neither climacteric nor non-climacteric fruit crop?
   a) Kiwifruit  b) Blackberry
   c) Raspberry  d) Blueberry
8. Which of the following fruit crops is associated with the 'Emblem of the United Nations'?
   a) Walnut  b) Olive
   c) Apple  d) Cherry

9. The finest variety of mandarin in the world is:
   a) Satsuma b) Kinnow
   c) Ponkan d) Coorg
10. Which of the following micronutrients is most widely deficient in Indian soils?
   a) Zn b) Fe
   c) B d) Cu
11. Marsh spot in pea occurs due to the deficiency of:
   a) Mg b) Mn
   c) B d) Fe
12. Pusa Srijan is an aneuploid rootstock of:
   a) Mango b) Apple
   c) Cherry d) Guava
13. The basic chromosome number (X=) of mango is:
   a) 10 b) 20
   c) 30 d) 40
14. Sporophytic system of self-incompatibility is found in:
   a) Apple b) Loquat
   c) Pineapple d) Mango
15. The term 'Auxin' is derived from:
   a) Latin word b) Greek word
   c) English word d) French word
16. Flavouring compound in citrus is:
   a) Hesperidin b) Limonin
   c) Citral d) Nootkatone
17. Which fruit crop is suitable for high pH:
   a) Fig b) Grape
   c) Guava d) Aonla
18. Where is the Central Coffee Research Institute located?
   a) Puttur (Karnatak b) Eluru (Andhra Pradesh)
   c) Pune (Maharashtr d) Chikamagalur (Karnataka)

19. Which crop is a good source of L-Dopa (Dopamine) to cure Parkinson disease in humans?
    a) Broccoli  b) Broad bean
    c) Celery  d) Lettuce
20. Where does pear primarily bear its fruit?
    a) Main stem  b) Roots
    c) Shoots  d) Spurs
21. Which of the following is an excellent autumn colour foliage?
    a) Codiaeum variegatum  b) Camellia japonica
    c) Acer palmatum  d) Polyanthes tuberosa
22. Which of the following crops is cauliforous?
    a) Persimmon  b) Peach
    c) Citrus  d) Jackfruit
23. Name the most suitable clonal root stock for replant situation of apple.
    a) Merton-793  b) MM-111
    c) MM-106  d) MM-104
24. In which crop is Gynoecy sex commercially exploited for hybrid seed production:
    a) Watermelon  b) Cucumber
    c) Ridge gourd  d) Muskmelon
25. Name the onion variety which is picking type and suitable for export:
    a) Agrifound Dark Red  b) Agrifound Light Red
    c) Agrifound Red  d) Agrifound Rose
26. Name the enzyme that causes 'Browning' in apple:
    a) Hydrogenase  b) Phenolase
    c) Polyphenol oxidase  d) Esterase
27. Which of the botanical varieties of the lettuce has excellent shipping and handling abilities?
    a) Butter head type  b) Crips head type
    c) Cos type  d) Celery lettuce
28. Which variety of banana is used for making baby food?
    a) Kunnan  b) Nendran
    c) Monthan  d) Hill banana

29. Which of the following has non-endospermic seed?
    a) Bael b) Pear
    c) Apple d) Ber
30. Which method can be employed to maintain self-incompatible inbred lines?
    a) Bud pollination b) Open pollination
    c) Cross pollination d) Compatible pollination
31. Which fatty acid is present in walnut?
    a) Lignan b) Omega 3
    c) Carboxylic acid d) Omega 6
32. Which of the following does not belong to genus Jasmnum?
    a) Italian jasmine b) Night jasmine
    c) Arabian jasmine d) Spanish jasmine
33. Name the fruit crop suitable for multi-storeyed cropping?
    a) Avocado b) Fig
    c) Phalsa d) Wood Apple
34. Which fruit crop is known as 'queen of nuts'?
    a) Hazel nut b) Pecan nut
    c) Pistachio nut d) Cashew nut
35. Which type of sex form is present in onion?
    a) Protandry b) Protogyny
    c) Chasmogamy d) Cleistogamy
36. 'Scooping' is an important operation in raising seed crop in which of the following vegetable crops:
    a) Broccoli b) Brussels sprouts
    c) Cabbage d) Cauliflower
37. Which of the following has large sized, white and fragrant flowers?
    a) Nyctanthes b) Magnolia
    c) Flame of forest d) Jacaranda
38. Core intact is one of the seed production methods used in which of the following crops:
    a) Broccoli b) Cabbage
    c) Knol khol d) Brussels sprout

39. Edible part of Asparagus is:
    a) Roots b) Tender shoots
    c) Fruits d) Flowers
40. T- bar system of training is followed in which fruit crop:
    a) Cherry b) Pear
    c) Kiwi fruit d) Pecan nut
41. How many agro-climatic zones are there in India?
    a) 5 b) 15
    c) 25 d) 30
42. In which state is rice grown below sea level:
    a) Tamil Nadu b) Karnataka
    c) Andhra Pradesh d) Kerala
43. Organic soils contains:
    a) >20% organic matter b) <20% organic matter
    c) >30% organic matter d) <30% organic matter
44. The family of carnation I 52 is ……
    a) Caryophyllaceae b) Pinaceae
    c) Iridaceae d) Compositae
45. Pusa Abhinav is the variety of:
    a) Acid lime b) Sweet lime
    c) Sweet orange d) Lemon
46. Time interval between irrigations is referred to as:
    a) Irrigation Efficiency b) Irrigation Frequency
    c) Irrigation Scheduling d) Irrigation Potential
47. Directorate of Floricultural Research is located at:
    a) Ahmedabad b) New Delhi
    c) Pune d) Firozabad
48. Which types of cucurbits are planted in polyhome?
    a) Andromonoecious b) Monoecious
    c) Gynoecium with parthenocarpy d) Gynoecious
49. The most favourable range of soil pH for tomato cultivation is:
    a) 4.5-5.5 b) 7.5-8.5
    c) 6.0-7.0 d) 7.2-8.0

50. Pusa Trissar is a variety of:
    a) Grape b) Guava
    c) Lime d) Sweet orange
51. Parents of Pusa Lalima are:
    a) Amrapali x Lal Sundari b) Amrapali x Sensation
    c) Amrapali x Janardan Pasand d) Dushehari x Sensation
52. Tagets patula marigold belongs to:
    a) African marigold b) Sinnet marigold
    c) French marigold d) Iris Lace
53. Fruit cracking in citrus occurs due to the deficiency of:
    a) Mg b) S
    c) Ca d) N
54. Rose hips of which species contains very high vitamin C:
    a) Rosa pimpinellifolia b) Rosa canina
    c) Rosa berberifolia d) All of these
55. Bull head is a physiological disorder of:
    a) Rose b) Gerbera
    c) Tuberose d) Anthurium
56. Ambri is an indigenous variety of:
    a) Apple b) Pear
    c) Mango d) Custard apple
57. Mimusops hexandra is the rootstock of which crop:
    a) Sapota b) Grape
    c) Peach d) Passion fruit
58. Which of the following citrus varieties is triploid?
    a) Tahiti lime b) Nova
    c) Page d) Redblush
59. The famous hanging garden is situated at:
    a) Delhi b) Pune
    c) Kanyakumari d) Mumbai
60. Xanthosoma sagittifollum is a:
    a) Root vegetable b) Bulb vegetable
    c) Tuber vegetable d) Stem vegetable

61. California Advance is the best pollinizer of:
    a) Apple b) Cherry
    c) Loquat d) European plum
62. Which of the following is a principal vector of plant viruses?
    a) White fly b) Aphid
    c) Jassid d) Leaf hooper
63. Little leaf of brinjal is transmitted by:
    a) Aphid b) Leaf hopper
    c) White fly d) Jassid
64. Pink colour of onion is caused due to:
    a) Anthocyanin b) Quercetin
    c) Carotene d) Delphinidin
65. Anticancer property of cabbage is due to the presence of:
    a) Sinigrin b) Indole-3-Carbinole
    c) Catechol d) Cheratin
66. Chiasmata become visible during the stage of prophase I of meiosis:
    a) Leptotene b) Diplotene
    c) Pachytene d) Zygotene
67. Which of the following is a male sterile variety of peach?
    a) J. H. Hale b) Florida Red
    c) Sharbati d) Red Haven
68. What is the threshold level of ethylene in fruits and vegetables?
    a) 0.01 mL/L b) 0.02 mL/L
    c) 0.03 mL/L d) 0.04 mL/L
69. Sauerkraut is prepared from:
    a) Cauliflower b) Cabbage
    c) Black carrot d) Cucumber
70. Pectin is basically a polymer of:
    a) Amino acids b) D-galactocurnic acid
    c) Gallic acid d) None of these
71. Which of the following is responsible for yellow colour in papaya?
    a) Carotene b) Xanthophyll
    c) Caricaxanthin d) Anthocyanin

72. Collection of falling nut in cashew is known as:
    a) Gelling b) Gleaning
    c) Ginning d) None of these
73. Which of the following varieties revalorised the pomegranate production in+ India?
    a) Bhagwa b) Ganesh
    c) Mridula d) Phule Arkta
74. Female flower mature before male flower in pecan nut is known as:
    a) Protoandry b) Protogyny
    c) Gynoecy d) Androecy
75. In India, CTV (citrus tristeza virus) is introduced from:
    a) Sri Lanka b) France
    c) USA d) Australia
76. Which of the following is a monoembyonic species of citrus?
    a) Citrus parodist b) Citrus maxima
    c) Citrus reticulate d) Citrus limon
77. Which of the following is subtropical evergreen pome fruit?
    a) Litchi b) Peach
    c) Plum d) Loquat
78. In India, apple was first introduced at:
    a) Shimla b) Kodaikanal
    c) Mashobra d) Ooty
79. Pinching of male flower sis effective in hybrid seed production of:
    a) Bottle gourd b) Ash gourd
    c) Bitter gourd d) Pointed gourd
80. Which of the following is highly tolerant to salt:
    a) Ash gourd b) Bitter gourd
    c) Bottle gourd d) Pumpkin
81. The red colour in chilli at ripening stage is due to ….. pigment:
    a) Capsaicin b) Capsanthin
    c) Lycopene d) Anthocyanin

82. Pusa Early Bunching is a variety of:
    a) Coriander b) Fenugreek
    c) Cumin d) Amaranthus
83. The most intensive type of vegetable farming is:
    a) Kitchen gardening b) Market gardening
    c) Truck gardening d) Floating gardening
84. Ploidy level of sweet potato is:
    a) Diploid b) Tetraploid
    c) Hexaploid d) Octaploid
85. Water melon bud necrosis is caused by:
    a) Fungus b) Bacteria
    c) Virus d) Mycoplasma
86. Cassava/Tapioca belongs to .........family:
    a) Araceae b) Dioscoreaceae
    c) Euphorbiaceae d) Asteraceae
87. Which of the following is a synthetic variety of cabbage?
    a) Pusa Mukta b) Pusa Sambandh
    c) Pusa Ageti d) Pusa Drum Head
88. Pusa Madhvi variety of onion is suitable for cultivation in:
    a) Kharif season b) Rabi season
    c) Zaid season d) Both Kharif and Rabi season
89. Brown anther male sterility is found in:
    a) Tomato b) Carrot
    c) Onion d) Cauliflower
90. Early Grano variety of onion was introduced from:
    a) USA b) Philippines
    c) UK d) Holland
91. Which of the following is a cross pollinated bean?
    a) Phaseolus vulgaris b) Phaseolus coccineus
    c) Phaseolus lanatus d) Phaseolus acutifolius
92. Which of the following is an acid tolerant vegetable?
    a) Sweet potato b) Okra
    c) Brinjal d) Potato

93. Which type of dichogamy is present in carrot?
    a) Protoandry b) Protogyny
    c) Both d) None of these
94. Non parasitic physiological disorder that occurs due to Ca deficiency in tomato is:
    a) Cat face b) Fruit cracking
    c) BER d) Puffiness
95. Which of the following sex form is found in muskmelon?
    a) Monoecius b) Dioecious
    c) Andromonoecious d) Gyoecious
96. Critical day length required for bulbing in garlic is:
    a) 8 hrs. b) 10 hrs.
    c) 12 hrs. d) 14 hrs.
97. The is done to obtain virus free plants:
    a) Another culture b) Embryo rescue
    c) Cell culture d) Meristem culture
98. Haploid plants can be obtained through:
    a) Anther culture b) Meristem culture
    c) Ovule culture d) Ovary culture
99. Which of the following hormone increases drought tolerance?
    a) IAA b) ABA
    c) Ethylene d) GA3
100. Which of the following crops contains highest cellulose?
    a) Cotton b) Sugarcane
    c) Jute d) Castor
101. Which of the following seed requires moisture for viability?
    a) Orthodox seed b) Recalcitrant seed
    c) Dormant seed d) Nonviable seed
102. Which of the following is used to control soil born pathogen in organic agriculture?
    a) Vermi compost b) Trichoderma
    c) FYM I 110. d) Oil cakes

103. Sesamum phyllody is caused by:

| | | | |
|---|---|---|---|
| a) | Virus | b) | Viroids |
| c) | Virforis | d) | Mycoplasma |

104. Golden Revolution is related to:

| | | | |
|---|---|---|---|
| a) | Dairy | b) | Fisheries |
| c) | Agriculture | d) | Horticulture |

105 is the site of dark reaction of photosynthesis:

| | | | |
|---|---|---|---|
| a) | Grana | b) | Stroma |
| c) | Vacuole | d) | Mitochondria |

106. Tetrad formation occurs during:

| | | | |
|---|---|---|---|
| a) | Leptotene | b) | Diakininsis |
| c) | Pachytene | d) | Zygotene |

107. Which of the following mechanism exists in avocado?

| | | | |
|---|---|---|---|
| a) | Heterodichogamy | b) | Dudichogamy |
| c) | Protogyny | d) | PDSD |

108. Radio, TV etc. are the examples of:

| | | | |
|---|---|---|---|
| a) | Mass communication | b) | Group communication |
| c) | Individual communication | d) | All of these |

109. 'Seed Order' came into Control existence in:

| | | | |
|---|---|---|---|
| a) | 1973 | b) | 1983 |
| c) | 1987 | d) | 1993 |

110. The tag colour of breeder seed is:

| | | | |
|---|---|---|---|
| a) | Golden yellow | b) | White |
| c) | Blue | d) | Opal green |

111. Thimann (1948) suggested the use the term for hormones of plants:

| | | | |
|---|---|---|---|
| a) | Plant growth regulators | b) | Phytohormone |
| c) | Plant growth hormones | d) | Plant growth retardant |

112. The response of plants to the relative length of day and night is known as:

| | | | |
|---|---|---|---|
| a) | Thermoperiodism | b) | Vernalization |
| c) | Photoperiodism | d) | Photorespiration |

113. The process of formation of mRNA from DNA is called:

| | | | |
|---|---|---|---|
| a) | Transcription | b) | Translation |
| c) | Replication | d) | Termination |

114. Binomial system of nomenclature of Linnaeus is based on:
   a) Historia plantarus  b) Genera plantarum
   c) An Enquiry into Plants  d) Origin of species

115. The growing of two or more than two crops simultaneously with a definite row arrangement in the same field is known as:
   a) Mixed cropping  b) Alley cropping
   c) Inter cropping  d) Relay cropping

116. Which of the following is an example of CAM plant?
   a) Banana  b) Dragon fruit
   c) Date palm  d) Pineapple

117. Rose belongs to .................family:
   a) Rutaceae  b) Rosaceae
   c) Compositeae  d) Liliaceae

118. 'Rashtrapati Bhawan' garden at New Delhi was laid out by:
   a) Shahjahan  b) Lady Harding
   c) Lord Curzon  d) Edwin Lutyens

119. Inflorescence of anthurium is called:
   a) Spadix  b) Sorosis
   c) Hypanthodium  d) None of these

120. Ornamental value of bougainvillea lies in:
   a) Flower  b) Fruits
   c) Foliage  d) Colourful bract

121. 'Chitra' is an important cultivar of:
   a) Rose  b) Bougainvillea
   c) Tulip  d) Tuberose

122. Which of the following is an Indian origin annual?
   a) Zinnia  b) Portulaca
   c) Balsam  d) Aster

123. Which of these flowers is day neutral?
   a) Companula  b) Rudbeckia
   c) Chrysanthemum  d) Gomphrena

124. Pinching is generally done in quality production of:

| | | | |
|---|---|---|---|
| a) | Carnation | b) | Dahlia |
| c) | Rose | d) | Aster |

125. Bulbophyllum is an important ................:

| | | | |
|---|---|---|---|
| a) | House plant | b) | Palm |
| c) | Orchid | d) | Succulent |

126. Bluing of rose petals is due to:

| | | | |
|---|---|---|---|
| a) | Accumulation of ammonia | b) | More salt |
| c) | Less sugar | d) | Bacteria |

127. Cassia fistula produces..............................flowers:

| | | | |
|---|---|---|---|
| a) | Red | b) | Blue |
| c) | Yellow | d) | White |

128. Kamrupa is a recently released variety of:

| | | | |
|---|---|---|---|
| a) | Arecanut | b) | Coconut |
| c) | Cashew nut | d) | Cocoa |

129. CAM plant is ............:

| | | | |
|---|---|---|---|
| a) | Flood tolerant | b) | Drought tolerant |
| c) | Acid tolerant | d) | Salt tolerant |

130. Segmental polyploidy is found in which of the following vegetable crops?

| | | | |
|---|---|---|---|
| a) | Potato | b) | Onion |
| c) | Radish | d) | Cauliflower |

131. A chemical used as a preservative for coloured fruit is:

| | | | |
|---|---|---|---|
| a) | Sodium benzoate | b) | KMS |
| c) | Sugar | d) | Salt |

132. R-enamel cans are used for packing of:

| | | | |
|---|---|---|---|
| a) | Acid food | b) | Non-acid food |
| c) | Both | d) | None of these |

133. Fruit cracking in pomegranate occurs due to the deficiency of:

| | | | |
|---|---|---|---|
| a) | Zn | b) | Mg |
| c) | Mn | d) | B |

134. Which of the following species is not used as rootstock for rose?

| | | | |
|---|---|---|---|
| a) | Rosa indica | b) | Rosa multiflora |
| c) | Rosa damascena | d) | Rosa bourborniana |

135. Which of the following crop is stored in cold storage in India?

a) Potato b) Onion
c) Garlic d) Carrot

136. Which of the following crop contains highest polyphenols?

a) Apple b) Blueberry
c) Pepper d) Beet root

137. The process in which a gene of interest is located and copied out of DNA extracted from an organism is known as:

a) Gene cloning b) Gene mapping
c) Gene knockout d) All of these

138. Gerbera was named in honour of:

a) German botanist b) French botanist
c) African botanist d) American botanist

139. Loquat belongs to ..........family:

a) Rutaceae b) Lauraceae
c) Rosaceae d) Ebenaceae

140. Economic part of anthurium is:

a) Spathe b) Flower
c) Bracts d) None of these

141. Chromosome number of modern dahlia is ..........:

a) 16 b) 32
c) 48 d) 64

142. Oelonix regia is commonly known as:

a) Bottle brush b) Gulmohar
c) Palas d) Jacaranda

143. Which of the following flower is a major source of carotenoids and lutein?.

a) African marigold b) Chrysanthemum
c) Jasmine d) Dahlia

144. Which of the following is a pink bract cultivar of bougainvillea?

a) Shubra b) Crimson Red
c) James Walker d) Formosa

145. Which of the following is not a variety of chrysanthemum?

a) Rakhee b) Indira
c) Keerti d) Kiran

146. 'Pride of India' is the common name of:

| a) Queen's flower | b) Balam khira |
|---|---|
| c) Barachampa | d) None of these |

147. Spores are used to propagate:

| a) Mosses | b) Ferns |
|---|---|
| c) Asparagus | d) Bromeliads |

148. The plant known as sweetener is:

| a) Rose | b) Jasmine |
|---|---|
| c) Aloe | d) Stevia |

149. Gynoecious variety of papaya is:

| a) Pusa Giant | b) Pusa Dwarf |
|---|---|
| c) Pusa Majesty | d) Pusa Nanha |

150. Calcium deficiency symptoms first appear at:

| a) Older leaves | b) Younger leaves |
|---|---|
| c) Terminal bud | d) All of these |

151. Match the following:

| i) Pomegranate | a) Catkin |
|---|---|
| ii) Pecan nut | b) Panicle |
| iii) Coconut | c) Hypanthodium |
| iv) Litchi | d) Corymb |
| v) Pear | e) Spadix |

152. Match the following:

| i) Scab | a) Pear |
|---|---|
| ii) Heart rot | b) Mango |
| iii) Anthracnose | c) Grape |
| iv) Malformation | c) Apple |
| v) Fire blight | e) Pineapple |

153. Match the following:

| i) Pusa Srijan | a) Guava |
|---|---|
| ii) 99-R | b) Plum |
| iii) 13/1 | c) Cherry |
| iv) F12/1 | c) Mango |
| v) GF series | e) Grape |

154. Match the following:

| | | | |
|---|---|---|---|
| i) | Asparagus | a) | Petiole |
| ii) | Celery | b) | Root |
| iii) | Rhubarb | c) | Stem |
| iv) | Knol khol | c) | Seed |
| v) | Sweet potato | e) | Spears |

155. Match the following:

| | | | |
|---|---|---|---|
| i) | Night blindness | a) | Vitamin B, |
| ii) | Beriberi | b) | Vitamin A |
| iii) | Scurvy | c) | Vitamin $B_3$ |
| iv) | Pellagra | d) | Vitamin C |
| v) | Sterility | e) | Vitamin E |

156. Match the following:

| | | | |
|---|---|---|---|
| i) | Pusa Madhvi | a) | Onion |
| ii) | Pant Anupama | b) | French bean |
| iii) | Hisar Lalit | c) | Tomato |
| iv) | Pusa Kesar | c) | Chilli |
| v) | CH-1 | e) | Carrot |

157. Match the following:

| | | | |
|---|---|---|---|
| i) | Pinjore garden | a) | Rashtrapati Bhawan garden |
| ii) | Japanese garden | b) | Ethylene inhibitor |
| iii) | Mughal garden | c) | Chandigarh |
| iv) | 1-MCP | c) | Buddha jayantii garden |
| v) | Preservative | e) | Potassium meta bisulphite |

158. Match the following:

| | | | |
|---|---|---|---|
| i) | Kesar | a) | Mandarin |
| ii) | Washington Navel | b) | Sweet orange |
| iii) | Kinnow | c) | Walnut |
| iv) | Roopa | c) | Pear |
| v) | Bartlett | e) | Mango |

159. Match the following:

| | | | |
|---|---|---|---|
| i) | Sweet potato | a) | 48 |
| ii) | Potato | b) | 90 |
| iii) | Brinjal | c) | 24 |
| iv) | Pea | c) | 18 |
| v) | Cabbage | e) | 14 |

160. Match the following:

| | | | |
|---|---|---|---|
| i) | Dog flower | a) | Cactus |
| ii) | Kamini | b) | Climber |
| iii) | Jacaranda | c) | Tree |
| iv) | Duck flower | c) | Shrub |
| v) | Bird nest | e) | Annual |

## Answers Key

| | | | | | | | | | | | | | |
|---|---|---|---|---|---|---|---|---|---|---|---|---|---|
| 1. | (b) | 2. | (a) | 3. | (a) | 4. | (a) | 5. | (a) | 6. | (c) | 7. | (c) |
| 8. | (b) | 9. | (c) | 10. | (a) | 11. | (b) | 12. | (c) | 13. | (b) | 14. | (c) |
| 15. | (b) | 16. | (a) | 17. | (c) | 18. | (c) | 19. | (a) | 20. | (c) | 21. | (c) |
| 22. | (c) | 23. | (a) | 24. | (b) | 25. | (c) | 26. | (c) | 27. | (b) | 28. | (a) |
| 29. | (c) | 30. | (a) | 31. | (b) | 32. | (b) | 33. | (c) | 34. | (b) | 35. | (a) |
| 36. | (c) | 37. | (b) | 38. | (b) | 39. | (b) | 40. | (c) | 41. | (b) | 42. | (c) |
| 43. | (a) | 44. | (a) | 45. | (a) | 46. | (c) | 47. | (c) | 48. | (c) | 49. | (c) |
| 50. | (a) | 51. | (c) | 52. | (c) | 53. | (c) | 54. | (b) | 55. | (a) | 56. | (a) |
| 57. | (a) | 58. | (a) | 59. | (c) | 60. | (c) | 61. | (c) | 62. | (b) | 63. | (b) |
| 64. | (b) | 65. | (b) | 66. | (b) | 67. | (a) | 68. | (a) | 69. | (b) | 70. | (b) |
| 71. | (c) | 72. | (b) | 73. | (b) | 74. | (b) | 75. | (c) | 76. | (b) | 77. | (c) |
| 78. | (c) | 79. | (a) | 80. | (a) | 81. | (b) | 82. | (b) | 83. | (b) | 84. | (c) |
| 85. | (c) | 86. | (c) | 87. | (b) | 88. | (d) | 89. | (b) | 90. | (a) | 91. | (b) |
| 92. | (c) | 93. | (a) | 94. | (c) | 95. | (c) | 96. | (c) | 97. | (c) | 98. | (a) |
| 99. | (b) | 100. | (a) | 101. | (b) | 102. | (b) | 103. | (c) | 104. | (d) | 105. | (b) |
| 106. | (c) | 107. | (c) | 108. | (a) | 109. | (b) | 110. | (a) | 111. | (b) | 112. | (c) |
| 113. | (a) | 114. | (b) | 115. | (c) | 116. | (c) | 117. | (b) | 118. | (c) | 119. | (a) |
| 120. | (d) | 121. | (b) | 122. | (c) | 123. | (c) | 124. | (a) | 125. | (c) | 126. | (a) |
| 127. | (c) | 128. | (b) | 129. | (b) | 130. | (a) | 131. | (a) | 132. | (a) | 133. | (c) |
| 134. | (c) | 135. | (a) | 136. | (a) | 137. | (a) | 138. | (a) | 139. | (c) | 140. | (a) |
| 141. | (b) | 142. | (b) | 143. | (a) | 144. | (c) | 145. | (c) | 146. | (a) | 147. | (b) |
| 148. | (c) | 149. | (c) | 150. | (c) | | | | | | | | |

| Q. | 151. | 152. | 153. | 154. | 155. | 156. | 157. | 158. | 159. | 160. |
|---|---|---|---|---|---|---|---|---|---|---|
| i) | (c) | (c) | (a) | (e) | (b) | (a) | (c) | (e) | (b) | (e) |
| ii) | (a) | (e) | (e) | (c) | (a) | (b) | (c) | (b) | (a) | (d) |
| iii) | (e) | (c) | (c) | (a) | (c) | (c) | (a) | (a) | (c) | (c) |
| iv) | (b) | (b) | (c) | (c) | (c) | (e) | (b) | (c) | (e) | (b) |
| v) | (c) | (a) | (b) | (b) | (e) | (c) | (e) | (c) | (d) | (a) |

# 8

# ICAR – JRF Horticulture Exam – 2022

1. Which portion of the sugarcane is the best for use as seed?

   a) Top 1/3 to 1/2 portion  b) Top 2/3 to 3/4 portion

   c) 2/3 portion of bottom  d) 3/4 portion of bottom

2. Flower drop and less fruit setting/fruit drop in chilli can be controlled by spraying

   a) Sulphur (1.52%) at 10 days intervals at growing stage.

   b) Planofix (NAA) @ 2-3 ml per 10 litres water at flowering stage.

   c) Dicofol (0.25%) at flowering stage.

   d) Fipronil (5% SC) 1ml per litre at flowering stage.

3. Mango hybrid Pusa Peetamber is a cross between

   a) Ratna × Alphons  b) Neelam × Alphonso

   c) Alphonso × Banganpalli  d) Amrapali × Lal Sundari

4. Which of the following is a biological agent for supperssing Parathenium hysterophorus?

   a) Mexican beetle  b) Gall fly

   c) Grass hopper  d) Tricoderma

5. The water requirement of wheat is

   a) 20-25 cm  b) 30-35 cm

   c) 50-55 cm  d) 60-65 cm

6. How many months after planting the ginger crop is ready for harvesting? Options:

   a) 14-16  b) 4-5

   c) 11-12  d) 7-9

7. Which of the following is NOT true for cashew nut shell liquid?
   a) Naturally occurring phenol
   b) Used in formulation of resins, detergents, insecticides and dyes
   c) It is obtained during processing of nuts in isolation of kernels
   d) Naturally occurring tannins
8. Zinc is responsible for the biosynthesis of which plant auxin ?
   a) NAA b) 2,4-D
   c) IBA d) IAA
9. Which of the following gladiolus cultivars were developed through spontaneous mutation?
   a) Pusa Manmohak b) Salman's Sensation
   c) Ratna's Butterfly d) Gunjan

   Choose the correct answer from the options given below:
   a) A and B only b) A and C only
   c) B and C only d) A and D only
10. Given below are two statements, one is labelled as Assertion A and the other is labelled as Reason R

    Assertion A: When black pepper is grown on slopes, due care has to be taken to avoid slopes facing south and southwest.

    Reason R: The northern and northern eastern slopes planting prevent the scorching effect of sun during summer.

    In light of the above statements, choose the most appropriate answer from the options given below
    a) Both A and R are correct and R is the correct explanation of A
    b) Both A and R are correct but R is NOT the correct explanation of A
    c) A is correct but R is not correct
    d) A is not correct but R is correct
11. According to planning commission (1985-1990) India is divided into how many agro climatic zone?
    a) 12 b) 15
    c) 20 d) 25

12. Make the correct sequence of steps involved in the pureline selection method of flower crop improvement?

    a) Conducting preliminary yield trials of progenies.

    b) Selection of individual plants from a local variety or some other genetically variable and homozygous population.

    c) Conducting replicated yield trials at several locations.

    d) Planting of progenies from individual plants separately with proper spacing.

    e) Release of best progeny as variety and seed multiplication for distribution.

    Choose the correct answer from the options given below

    a) A, B, D, E, C, b) A, D, B, C, E.

    c) B, D, A, C, E. d) D, E, A, C, B

13. Match List I with List II

| List I | | List II | |
|---|---|---|---|
| Book | | Author | |
| A. | Fundamentals of Horticulture | I. | B.S. Chudawat |
| B. | Fruit breeding | II. | M.R. Dinesh |
| C. | Fruit Production in India | III. | Jitender Singh |
| D. | DrylandHorticulture | IV. | W.S. Dhillon |
| | | V. | Amar Singh |

Choose the correct answer from the options given below:

a) A-III, B-II, C - IV, D – I b) A-I, B-II, C-III, D - IV

c) A - II, B-III, C - V, D - IV d) A - III, B-II, C-IV, DI,

14. Match List I with List II

| List I | | List II | |
|---|---|---|---|
| Garden Name | | Purpose | |
| A. | Pramododyan | I. | Dedicated to lord Krishna |
| B. | Udyan | II. | Enjoyment of the royal couples |
| C. | Brishavatika | III. | Playing chess, enjoying dance of maids, etc. by the king. |
| D. | Nandanavana | IV. | Enjoying life with courtesans by the high placed persons in the king's court |

Choose the correct answer from the options given below:

a) A-III, B-IV, C-II, D-I　　b) A-I, B-II, C-IV, D-III

c) A-II, B-III, C-IV, D-I　　d) A-IV, B-III, C-I, D-II

15. Chilling requirement for plum (Japanese)

a) 200-300 hours　　b) 400-500 hours

c) 700-1000 hours　　d) 1200-1500 hours

16. Which of the following climber possesses special structure-tendrils to climb over a support?

a) Bougainvillea peruviana　　b) Ficus repens,

c) Quisqualis indica　　d) Antigonon leptopus

17. Given below are two statements, one is labelled as Assertion A and the other is labelled as Reason R

Assertion A: Gametophytic self incompatibility was first described by Hughes and Babcock in Crepis foetida.

Reason R: In gametophytic system, incompatibility reaction of pollen is determined by its own genotype and not by the genotype of the plant on which it is produced.

In light of the above statements, choose the most appropriate answer from the options given below

a) Both A and R are correct and R is the correct explanation of A

b) Both A and R are correct but R is NOT the correct explanation of A

c) A is not correct but R is correct

18. Sequence the Agri export zone according to the fruit.

a) Tripura and West Bengal　　b) Maharashtra & Gujarat

c) Maharashtra　　d) Chitoor

e) Tamil Nadu

Choose the correct answer from the options given below Options:

a) Kesar Mango-Mango Pulp-Grape-Cashewnut-Pineapple,

b) Pineapple-Kesar Mango-Grape-Mango Pulp-Cashewnut

c) Pineapple-Cashewnut-Mango Pulp-Kesar Mango-Grape

d) Grape-Pineapple-Kesar Mango-Cashewnut-Mango Pulp

19. Which of the following is NOT true for Ney Poovan banana?

a) It is AB type banana　　b) Diploid variety

c) Highly susceptible BBMV　　d) Resistant to nematodes

20. Given below are two statements, one is labelled as Assertion A and the other is labelled as Reason R

Assertion A: Deformed flowers and browning of throat are major disorders of orchids.

Reason R: These disorders of orchids occur due to low temperature and chilling injury.

In light of the above statements, choose the most appropriate answer from the options given below

a) Both A and R are correct but R is NOT the correct explanation of A

b) Both A and R are correct and R is the correct explanation of A

c) A is correct but R is not correct

d) A is not correct but R is correct

21. Match List I with List II

| List I | List II |
|---|---|
| A. Kanji | I. Grapes |
| B. Cider | II. Carrot |
| C. Wine | III. Apple |
| D. Marmalade | IV. Citrus |
| | V. Pear |

Choose the correct answer from the options given below:

a) A-II, B-III, C-I, D-IV b) A-II, B-IV, C-I, D-V

c) A-II, B-III, C-IV, D-V d) A-III, B-II, C-I, D-IV

22. Sequence the disorder with their respective fruits

a) Soft nose b) Goose flesh

c) Stalk necrosis d) Granulation

e) Black Heart

Choose the correct answer from the options given below Options:

a) Banana-Citrus-Grape-Pineapple-Mango

b) Mango-Pineapple-Grape-Citrus-Banana

c) Pineapple-Citrus-Banana-Grape-Mango

d) Mango-Banana-Grape-Citrus-Pineapple

23. Given below are two statements

Statement I: Term heterosis was first used by Shull (1914) and is defined as the superiority of an F1 hybrid over both its parents in terms of yield and some other characters.

Statement II: Genetic basis of heterosis is explained by two main theories namely dominance hypothesis and over dominance hypothesis.

In light of the above statements, choose the most appropriate answer from the options given below

Options:

a) Both Statement I and Statement II are correct

b) Both Statement I and Statement II are incorrect

c) Statement I is correct but Statement II is incorrect

d) Statement I is incorrect but Statement II is correct

24. Which of the following colours belong to hard or warm colours in "Colour scheme"? Options:

a) Blue, violet and green
b) White, black and grey
c) Red, white and green
d) Red, orange and yellow

25. How much amount of urea would be required for supplying 5 kg of N? Options:

a) 5.25
b) 9.62
c) 10.87
d) 15.66

26. Which one of the following is NOT an example of mesophytes?

a) Wheat
b) Maize
c) Cotton
d) Low land rice

27. Which grapefruit cultivar has originated as a pink fleshed budsport of Marsh?

a) Duncan
b) Foster
c) Thompson
d) Star Ruby

28. Which of the following is the most common rootstock for budding in rose under North Indian conditions?

Options:

a) Rosa foetida
b) Rosa lucida
c) Rosa multiflora
d) Rosa indica var. odorata

29. Given below are two statements, one is labelled as Assertion A and the other is labelled as Reason R

Assertion A: High rainfall and long photoperiod promote vine growth and reduce the tuber yield of sweet potato.

Reason R: Temperature less than 25 °C was found critical for tuberization in sweet potato. In light of the above statements, choose the most appropriate answer from the options given below

Options:

a) Both A and R are correct and R is the correct explanation of A

b) Both A and R are correct but R is NOT the correct explanation of A

c) A is correct but R is not correct

d) A is not correct but R is correct

30. Match List I with List II

| List I | List II |
|---|---|
| Vegetable Hybrid | Used genetic system |
| A. Pusa Nayanjyoti | I. GMS |
| B. Pusa Hybrid-2 | II. CMS |
| C. PusaSeedless Cucumber-6 | III. SelfIncompatibility |
| D. CH-1 | IV. Gynoecy |

Choose the correct answer from the options given below: Options:-

a) A-II, BI, C-III, D-IV
b) A-IV, B-II, CI, D-III
c) A-III, B-II, C-IV, D-I
d) A-II, B-III, C-IV, D-I

31. Spice Board India' works under administrative control of

a) Ministry of Agriculture and Farmer Welfare

b) Ministry of Science and Technology

c) Ministry of Food Processing Industries

d) Ministry of Commerce and Industry

32. Which of the following is NOT a F1 hybrid of chilli? Options:

a) Kashi Tej
b) Kashi Ratna
c) Kashi Surkh
d) Kashi Sinduri

33. Which of the following new curcuminoid has been isolated from nematocidally active fraction of turmeric?

a) Demethoxy curcumin
b) Diferuloymethane
c) Bisdemethoxycurcumin
d) Cyclocurcumin

34. Which of the following shrubs have white flowers?

a) Ixora acuminata
b) Ixora armeniaca
c) Ixora superba
d) Ixora chinensis

Choose the correct answer from the options given below:

Options:

a) B and D only
b) A and B only
c) A and C only
d) A and D only

35. Given below are two statements, one is labelled as Assertion A and the other is labelled as Reason R

Assertion A: Guava is ideal fruit for jelly preparation

Reason R: Guava is rich in pectin and acids

In light of the above statements, choose the most appropriate answer from the options given below

a) Both A and R are correct and R is the correct explanation of A
b) Both A and R are correct but R is NOT the correct explanation of A
c) A is correct but R is not correct
d) A is not correct but R is correct

36. Match List I with List II

| List-I | List-II |
|---|---|
| Planting System | Features |
| A. Hexagonal | I. Filler Plant |
| B. Quincunx | II. Originated from Israel |
| C. Medow orcharding | III. 15% more plant than square system |
| D. Contour | IV. Plant spacingvaries |

Choose the correct answer from the options given below: Options:

a) A-III, BI, C-II, D-IV
b) A-I, B-II, C-III, D-IV
c) A-III B-I, C-IV, D-II
d) A-IV, B-III, C-II, D-I

37. Which of the following part of carrot contains the highest carotene level?

a) Top part of root
b) Middle part of root
c) Distal end of root
d) Middle Core

38. Black heart/ hollow heart of potato is caused by

a) Low soil temperature (less than 13°C)
b) Favourable supply of oxygen during storage and transport

c) Rapid growth and development of tubers followed by sudden irrigation

d) It is more common in small sized tubers

39. Given below are two statements

Statement I: In plum, both self fruitful and self incompatible varieties are available Statement II: Green Gage is self unfruitful variety of plum

In light of the above statements, choose the correct answer from the options given below Options:

a) Both Statement I and Statement II are true

b) Both Statement I and Statement II are false

c) Statement I is true but Statement II is false

d) Statement I is false but Statement II is true

40. Which one of the following is an unfavourable factor for self pollination?

a) Cleistogamy b) Geitonogamy

c) Monoecy d) Protogyny

41. Which of the following is a thermo-photo insensitive variety of chrysanthemum?

a) Pusa Aaditya b) Birbal Sahni

c) Pusa Anmol d) Himanshu

42. The combined effect of alleo chemical and competition of weed with crop, is known as.

a) Interference b) Suppressed weed growth

c) Allelopathy d) Hindrance

43. Sugary secretion (gummosis) is a physiological disorder of Options:

a) Fennel b) Fenugreek

c) Cumin d) Clove

44. Agricultural Education Day is celebrated on

a) 1st December b) 3rd December

c) 5th September d) Guru Purnima

45. Which of the following pineapple variety belong to Cayenne group?

a) Kew b) Charlotte Rothchild

c) Jaldhup d) Lakhat

e) Giant Kew

Choose the correct answer from the options given below: Options:

a) B only b) A, B and E only

c) A only d) A and D only

46 Under north Indian condition plant crop of sugarcane should be harvested during which period to obtain profuse ratoon initiation

Options:

a) Mid December - Mid January b) Mid January- Mid February

c) Mid February - Mid March d) Mid March - Mid April

47. Which is NOT correct about 'Shellac coating" which is generally used for increasing the shelf life of vegetables?

a) Shellac is a resin secreted by the insect

b) Shellac is composed of aleuritic and shelloic acids

c) Shellac is incompatible with waxes,

d) Shellac is permitted as an indirect food additive

48. Given below are two statements, one is labelled as Assertion A and the other is labelled as Reason R

Assertion A: Ripe mango peels show colour variation from green to greenish yellow, yellow red to violet

Reason R: It is due to the concentrated action of pectin

In light of the above statements, choose the correct answer from the options given below

a) Both A and R are true and R is the correct explanation of A

b) Both A and R are true but R is NOT the correct explanation of

c) A is true but R is false

d) A is false but R is true

49. Given below are two statements

Statement I: Banana develops deep root system to trap water and nutrients from the lower profile of soil.

Statement II: Depth and drainage are therefore important factors while selecting soil for Banana

In light of the above statements, choose the most appropriate answer from the options given below

a) Both Statement I and Statement II are correct

b) Both Statement I and Statement II are incorrect

c) Statement I is correct but Statement II is incorrect, 2063

d) Statement I is incorrect but Statement II is correct

50. Which of the following mushroom is still dominating (more than 85% share) the Indian mushroom industery?

a) Agaricus bisporus b) Lentinula edodes

c) Volvariella volvacea d) Pleurotus eous

51. Young fruits are the site for synthesis of

a) IAA b) GA

c) Cytokinin d) Ethylene Answer

52. Which of the following wild cucumber has highest level of cucurbitacins?

a) *C. naudinianus* b) *C. hookei*

c) *C. foetidissima* d) *C. melo*

53. Which biological agent can control Parthenium hysterophorus weed?

a) Teleonima scruplosa b) Neochetina bruchi

c) Dactylopius tomentosus d) Zygogramma bicolorata

54. The seedling of which plants are thorny during juvenile phase

a) Apple b) Pear

c) Citrus d) Guava

e) Mango

Choose the correct answer from the options given below: Options:

a) A, B and D only b) A, B and C only

c) C only d) A and C only

55. Match List I with List II

| List-I | List-II |
|---|---|
| A. Litchi | I. Aril |
| B. Mango | II. Mesocarp and endocarp |
| C. Banana | III. Mesocarp |
| D. Grape | IV. Pericarp and Placenta |
| | V. Endosperm |

Choose the correct answer from the options given below:

a) A-I, B-III, C-II, D-IV b) A-I, B-III, C-II, D-V

c) A-I, B-II, C-III, D-IV d) A-I, B-IV, C-III, D-II

56. Which one of the following is a dioecious mutant of papaya?

a) Solo b) Pusa Nanha

c) Co 5 d) Co 7

57. Grape variety suitable for preparing Monukka Options:

a) Anabe-e-Shahi b) Bangalore Blue

c) Perlette d) Sharad Seedless

58. Given below are two statements

Statement I: Pusa Gaurav variety of tomato is suitable for long distance transportation. Statement II: Pusa Lal Meeruti variety of tomato is induced mutant by chemical mutagen In light of the above statements, choose the most appropriate answer from the options given below

Options:

a) Both Statement I and Statement II are correct

b) Both Statement I and Statement II are incorrect

c) Statement I is correct but Statement II is incorrect

d) Statement I is incorrect but Statement II is correct

59. Which of the following is specific origin of Bottlegourd?

a) Tropical East Africa b) Malabar region of Kerala

c) Canary islands d) Mediterranean

60. Match List I with List II for control of citrus pest through parasite/predators

| List I | List II |
|---|---|
| A. White Flies | I. Coccinellaseptempunctata |
| B. Mealy bug | II. Cryptolaemus montrouzieri |
| C. Citrus psylla | III. Brumus suturalis |
| D. Citrusleaf miner | IV. Citrospillusquadristriantus |

Choose the correct answer from the options given below:

Options:

a) A-I, B-II, C-III, D-IV b) A-IV, B-I, C-II, D-III

c) A-III, B-II, CI, D-IV d) A-II, B-IV, CI, D-III

61. Given below are two statements, one is labelled as Assertion A and the other is labelled as Reason R

Assertion A: Whip tail physiological dosorder is caused due to molybdenum deficiency in cauliflower.

Reason R: The occurance of whip tail disorder particularly in the soil pH above 5.5.

In light of the above statements, choose the most appropriate answer from the options given below

a) Both A and R are correct and R is the correct explanation of A

b) Both A and R are correct but R is NOT the correct explanation of A

c) A is correct but R is not correct

d) A is not correct but R is correct

62. Which of the following pigment is responsible for blue colour of flowers? Options:

a) Pelargonidin  b) Lycopene

b) Delphinidin  c) Lutein

63. Eggplant specific piquant and bitter taste is due to the presence of Options:

a) Flavonoids  b) Solasonine

c) Solanidine  d) Spirostanol

64. The tedious process of emasculation in pineapple is overcome due to: Options:

a) Incompatibility of pollen  b) Incompatibility of pollen and ovule

c) Incompatibility of ovule  d) Incompatibility of stigma

65. Match List I with List II

| List | | List II | |
|---|---|---|---|
| Characteristics of Onion | | Chemicals | |
| A. | Flavour and Pungency | I. | Anthocyanin |
| B. | Yellow coloured | II. | High dry matter |
| C. | Red coloured | III. | Allyl propyl disulphide |
| D. | White Coloured | IV. | Quercetic |

Choose the correct answer from the options given below:

a) A-IV, B-II, C-I, D-III  b) A-III, B-IV, C-I, D-II

c) A-I, B-III, C-II, D-IV  d) A-II, B-I, C-IV, D-III

66. Arrange the following steps of gladiolus hybrid development in correct sequence?

a) Development of hybrids

b) Combining ability testing of inbred lines

c) Development of inbred lines

d) Seed production of hybrids

e) Parents selection and hybrid development

Choose the correct answer from the options given below Options:

a) C, D, E, B, A b) A, D, C, B, E

c) C, B, E, A, D d) A, D, E, C, B

67. Match List I with List II

| List I | List II |
|---|---|
| Vegetable crop | Minimum soil temperature requirement for seed germination |
| A. Brinjal | 1. 1.6°C |
| B. Cauliflower | II. 10°C |
| C. Onion | III. 4.4°C |
| D. Palak | IV. 15.5°C |

Choose the correct answer from the options given below:

a) A-II, B-I, C-III, D-IV b) A-IV, B-III, CI, D-II

c) A-IV, B-II, C-III, D-I d) A-II, B-III, C-IV, D-I

68. Polyembryony is common in citrus except

a) Pumelo b) Tahti lime

c) Kinnow d) Sweet orange

e) Acid lime

Choose the correct answer from the options given below:

Options:

a) A, B and D only b) A only

c) B only d) A and B only

69. Pineapple variety that has got Geographical Indication Status Options:

a) Amrita b) Giant Kew

(c) Kew d) Vazhakulam

70. Stem rot disease of rice occurs at a stage from Options:

a) Panicle initiation to booting b) Pre tillering to mid tillering

c) Pre flowering-Post flowering d) Nursery to Tillering

71. Given below are two statements

Statement I: Male sterility is characterized by nonfunctional pollen grains, while female gametes function normally.

Statement II: Male sterility is found common in rose, ageratum and chrysanthemum flower crops.

In light of the above statements, choose the most appropriate answer from the options given below

a) Both Statement I and Statement II are correct

b) Both Statement I and Statement II are incorrect

c) Statement I is correct but Statement II is incorrect

d) Statement I is incorrect but Statement II is correct

72. What is the main purpose of blanching in cauliflower?

a) Increase in yield
b) To improve the nutrition
c) To keep the curd white
d) To keep the curd compact

73. The Indian group of radishes including the rat-tail evolved in

a) Northern western region
b) Northern eastern region
c) Southern eastern region
d) Southern western region

74. The quality of air at a given location is characterised by AQI in the range of Options:

a) 0-200
b) 0-300
c) 0-400
d) 0-500

75. Match List I with List II

| List I | List II |
|---|---|
| Variety | Pedigree/Parent |
| A. Varsha Uphar | I. Lam selection-1 × Parvani Kranti |
| B. Pusa Sawami | II. Mutant from pusaSawami |
| C. Prabhani Kranti | III. Pusa Sawami × A. manihot |
| D. Punjab -8 | IV. Pusa Makhmali × IC142 |

Choose the correct answer from the options given below: Options:

a) A-I, B-IV, C-III, D-II
b) A-IV, B-II, C-I, D-III
c) A-III, B-I, C-II, D-IV
d) A-II, B-III, C-IV, D-I

76. The group 'Picotees' belongs to which of the following commercial flower crop?

a) Tuberose b) Rose
c) Carnation d) Orchids

77. Given below are two statements

Statement I: A chimera is a mixture of tissues of genetically different constitutions on the same part of plant.

Statement II: The differences usually arise due to irregular meiosis.

In light of the above statements, choose the most appropriate answer from the options given below

a) Both Statement I and Statement II are correct
b) Both Statement I and Statement II are incorrect
c) Statement I is correct but Statement II is incorrect
d) Statement I is incorrect but Statement II is correct

78. The first stable product of $C_3$ pathway is

a) Phosphoglycerate b) Oxaloacetate
c) Pyruvic acid d) Malate dehydrogenase

79. Which of the following is correct for low chilling varieties of apple?

a) Chilling requirement is less than 800 hours below 7°C
b) Poor in dessert quality
c) Michal is low chilling apple variety
d) Excellent dessert quality

80. Number of bee hives required for effective pollination in case of strawberry Options:

a) 25 b) 8-10
c) 12-15 d) 25 or more

81. Given below are two statements

Statement I: Stooling is successful and economical method of propagation in case of guava

Statement II: The number of stools per plant increase with age of the plant and can be done for many years on the same plant

In light of the above statements, choose the correct answer from the options given below

a) Both Statement I and Statement II are true

b) Both Statement I and Statement II are false

c) Statement I is true but Statement II is false

d) Statement I is false but Statement II is true

82. Given below are two statements

Statement I: During 1972-73 there was a apple epidemic in Jammu and Kashmir Statement II: Commercial scab resistant variety for Jammu and Kashmir are Shalimar Apple I and Shalimar Apple II

In light of the above statements, choose the correct answer from the options given below:

a) Both Statement I and Statement II are true

b) Both Statement I and Statement II are false

c) Statement I is true but Statement II is false

d) Statement I is false but Statement II is true

83. Which of the following orchids have sympodial growth pattern?

a) Phalaenopsis b) Dendrobium

c) Habenaria d) Cymbidium

Choose the correct answer from the options given below: Options:

a) A and B only b) A and C only

c) A and D only d) B and D only

84. Annual water requirement of banana

a) 500-700 mm b) 800-1000 mm

c) 1200-2200 mm d) 2500-3000 mm

85. Famous blue coloured variety 'Applause' belongs to which of the following flower crop? Options:

a) Lilium longiflorum b) Rosa hybrida

c) Dianthus caryophyllus d) Tagetes erecta

86. Which of the following is a new scientific name of Polianthes tuberosa?

a) Agave americana b) Agave amica

c) Agave parviflora d) Agave angustifolia

87. Given below are two statements, one is labelled as Assertion A and the other is labelled as Reason R

Assertion A: In monocot plants Grafting is difficult with low percentage of takes as compared to dicots.

Reason R: Monocots have vascular bundle scattered throughout the stem as compared to dicots.

In light of the above statements, choose the correct answer from the options given below Options:

a) Both A and R are true and R is the correct explanation of A

b) Both A and R are true but R is NOT the correct explanation of A

c) A is true but R is false

d) A is false but R is true

88. Match List I with List II

| List I | List II |
|---|---|
| A. Heliophytes | I. Plants which growin open sunny situation |
| B. Sciophytes | II. Plants which grow in shade |
| C. Obligative sciohytes | III. Plants which always grow in shade |
| D. Obligative heliophytes | IV. Plants which always grow under sunny situation |

Choose the correct answer from the options given below: Options:

a) A-I, B-II, C-III, D-IV b) A-I, B-II, C-IV, D-III

c) A-I, B-IV, C-III, D-II d) A-IV, B-III, C-II, D-I

89. Male and female plants of spinach are identified on the basis of Options:

a) High leaf sugars, chlorophyll and carotenoids in female plants

b) High leaf sugars, chlorophyll and carotenoids in male plants

c) Red venal pigmentation in male plants

d) Downward cupping in female leaves

90. Which of the following turf grasses are used to develop a lawn in cooler areas of India?

a) Agrostis palustris b) Cynodon dactylon

c) Lolium perene d) Paspalum notatum

Choose the correct answer from the options given below: Options:

a) A and B only b) A and C only

c) B and D only d) C and D only

91. Match List I with List II

| List I | List II |
|---|---|
| Crop Basics | Chromosome number |
| A. Carnation | I. X = 9 |
| B. Marigold | II. X = 30 |
| C. Tuberose | III. X = 15 |
| D. Chrysanthemum | IV. X = 12 |

Choose the correct answer from the options given below: Options:

a) A-I, B-II, C-IV, D-III b) A-III, B-IV, C-II, D-I

c) A-IV, B-III, C-I, D-II d) A-II, B-III, C-IV, D-I

92. Which of the following gardens are laid out on the pattern of Mughal gardens?

a) Rashtrapati Bhavan garden, New Delhi

b) Buddha Jayati Park, New Delhi

c) Mandor Garden, Jodhpur

c) Nishat Bagh, Kashmir

Choose the correct answer from the options given below: Options:

a) A and C only b) A and D only

c) B and C only d) C and D only

93. The elements of garden and landscape designs are

a) Line b) Focalization

c) Rhythm d) Habit

Choose the correct answer from the options given below: Options:

a) A and C only b) A and B only

c) B and C only d) A and D only

94. Dr. Puskarnath is associated to Options:

a) Tomato breeding b) Potato breeding

c) Okra breeding d) Cucurbit breeding

95. The soil with more than 20 kg/ha phosphorous are rated as

a) Low in P b) Medium in P

c) High in P d) Very low in P

96. What is the family of Tapioca?

a) Dioscoreaceae b) Convolvulaceae

c) Araceae d) Euphorbiaceae

97. Which coloured pollinizer can be included in the delicious plantations of apple?

a) Gloster b) Spartan

c) Smoothy d) Summer Queen

Choose the correct answer from the options given below Options:

a) A, B and D only b) A, B and C only

c) B, C and D only d) A, C and D only

98. Perennial plants having erect and bushy growth and attain a height of 0.5 to 4 meters, are called as.

Options:-

a) Seasonal flowers b) Trees

c) Shrubs d) Climbers

99. Match List I with List II with the related term

| List I | | List II | |
|---|---|---|---|
| A. | Draining of latex from mango fruit | I. | Verasion |
| B. | On set of ripening in grape | II. | Verdelli |
| C. | Summerproduction of lemon under water stress | III. | Desapping |
| D. | Stage of inflorescence emergence in pinneapple | IV. | Red Heart |

Choose the correct answer from the options given below: Options:

a) A-I, B-II, C-III, D-IV b) A-IV, B-I, C-II, D-III

c) A-III, B-I, C-II, D-IV d) A-II, B-III, C-IV, D-I

100. Which of the following is a major onion producing state in India? Options:

a) Gujarat b) Karnataka

c) Maharashtra d) Madhya Pradesh

101. Given below are two statements

Statement I: Seeds of subtropical fruits like mango and subtropical fruits like citrus germinate immediately after extraction from the fruits under favourable conditions of moisture, temperature and aeration

Statement II: However, the seeds extracted from temperate fruits like apple, pacr and peach do not germinate immediatelty even the seed is viable and condition for germination are favourable

In light of the above statements, choose the correct answer from the options given below Options:

a) Both Statement I and Statement II are true

b) Both Statement I and Statement II are false

c) Statement I is true but Statement II is false

d) Statement I is false but Statement II is true

102. What is the main cause of appearance of gold colour flecks on tomato fruits? Options:

a) Low K: Ca ratio
b) High Mg content
c) High EC level
d) Low Boron content

103. Given below are two statements

Statement I: Vermiculite is heat expanded mica which is main ingredients in rooting medium.

Statement II: Perlite is heavy rock material and easily sedimented during mixing in growing medium.

In light of the above statements, choose the correct answer from the options given below Options:

a) Both Statement I and Statement II are true

b) Both Statement I and Statement II are false

c) Statement I is true but Statement II is false

d) Statement I is false but Statement II is true

104. The isolation distance for seed production in case of papaya through open pollination is Options:

a) 1 Km
b) 2 Km
c) 3 Km
d) 5 Km

105. Coconut hybrids

a) VHC 3
b) Chandra Kalpa
c) Chandra Sarkara
d) Kera Sree
e) Pratap

Choose the correct answer from the options given below:

Options:

a) A, B and D only b) A, C and D only

c) A and E only d) C, D and E only

106. Which of the following variety of onion is suitable for dehydration? Options:

a) Punjab-48 b) Udaipur-101

c) Udaipur-103 d) N-53

107. Which of the following trees are grown for their ornamental foliage?

a) Bombax malabaricum b) Cedrus deodara

c) Jacaranda mimosifolia d) Thuja orientalis

Choose the correct answer from the options given below:

a) B and C only b) B and D only

c) A and C only d) C and D only

108. Given below are two statements, one is labelled as Assertion A and the other is labelled as Reason R

Assertion A Male sterility is considered to be most important for commercial hybrid seed production of vegetable crops.

Reason R The male sterile lines eliminate the tedious emasculation procedure and make it possible to produce hybrid seed.

In light of the above statements, choose the correct answer from the options given below Options:-

a) Both A and R are true and R is the correct explanation of A

b) Both A and R are true but R is NOT the correct explanation of A

c) A is true but R is false

d) A is false but R is true

109. Which of the following is NOT a desirable trait for mechanical harvesting of pickling cucumber?

Options:

a) Small mature fruit size b) Parthenocarpy

c) Indeterminate habit d) Short inter-nodal length

110. Petrea volubilis, a light climber with beautiful flowers belongs to which of the following family?

a) Bignoniaceae b) Verbenaceae

c) Polygonaceae d) Convolvulaceae

111. Homeostasis that leads to adaptability and stability is NOT favourded by Options:

a) Self-pollination
b) Heterogeneity
c) Heterozygosity
d) Genetic polymorphism

112. Soils of 'b' horizon are dominated by

a) Mineral matter
b) Organic matter
c) Clay, Fe, Al and humus
d) Sand and Silt particles

113. Which of the following is NOT true for rootstock 13 - 1? Options:

a) It is mango rootstock
b) Developed from Israel
c) Salt tolerant
d) Dwarfing rootstock

114. High yielding varieties programme was initially started with

a) 100 districts in country during 1965-66
b) 100 districts in the country during 1970-71
c) All district of Punjab and Haryana during 1965-66
d) 1000 district in the country during 1965-66

115. Given below are two statements, one is labelled as Assertion A and the other is labelled as Reason R

Assertion A: Grape seeds are stratified to break dormancy to get higher percentage of germination

Reason R: Seeds contain high level of GA and low level of phenols.

In light of the above statements, choose the correct answer from the options given below Options:

a) Both A and R are true and R is the correct explanation of A
b) Both A and R are true but R is NOT the correct explanation of A
c) A is true but R is false
d) A is false but R is true

116. Given below are two statements

Statement I: Tea Mosquito Bug is a major pest of cashew

Statement II: Mid season/ late season flowering varieties such as Bhaskara are able to escape from the severity of this pest

In light of the above statements, choose the correct answer from the options given below Options:

a) Both Statement I and Statement II are true
b) Both Statement I and Statement II are false

c) Statement I is true but Statement II is false

d) Statement I is false but Statement II is true

117. Given below are two statements, one is labelled as Assertion A and the other is labelled as Reason R

Assertion A: ABA helps during water stress

Reason R: ABA has a role in stomatal closing

In light of the above statements, choose the correct answer from the options given below Options:

a) Both A and R are true and R is the correct explanation of A

b) Both A and R are true but R is NOT the correct explanation of A

c) A is true but R is false

d) A is false but R is true

118. Steps for synthetic breeding are as given below:

a) Testing of specific combining ability effects

b) Growing inbred lines in isolation and allowing random mating

c) Development of inbred lines

d) Selection of about 5 inbred lines basis of higher sca effects

e) Testing performance of synthetics over varieties

Choose the correct answer from the options given below Options:

a) D, A, C, E, B
b) A, C, B, D, E

c) C, A, D, B, E
d) A, B, C, D, E

119. Mid season pear cultivars for high hills of Himachal Pradesh are

a) Bartlett
b) Starkrimson

c) Flemish Beauty
d) Conference

Choose the correct answer from the options given below: Options:

a) A, B and D only
b) A, B and C only

c) B, C and D only
d) A, C and D only

120. Arrange the following according establishment year of Institute/Organization

a) Indian Institute of Vegetable Research

b) Indian Institute of Horticultural Research

c) Indian Society of Vegetable Science

d) The Horticultural Society of India

e) Indian Council of Agriculrural Research

Choose the correct answer from the options given below Options:

a) D, E, B, C, A  b) E, D, C, A, B

c) E, D, B, C, A  d) D, E, A, B, C

## Answer Key

| | | | | | | |
|---|---|---|---|---|---|---|
| 1. (a) | 2. (b) | 3. (d) | 4. (a) | 5. (c) | 6. (d) | 7. (d) |
| 8. (d) | 9. (c) | 10. (a) | 11. (b) | 12. (c) | 13. (a) | 14. (c) |
| 15. (c) | 16. (d) | 17. (c) | 18. (b) | 19. (d) | 20. (b) | 21. (a) |
| 22. (d) | 23. (a) | 24. (d) | 25. (c) | 26. (d) | 27. (c) | 28. (d) |
| 29. (c) | 30. (d) | 31. (d) | 32. (d) | 33. (d) | 34. (c) | 35. (a) |
| 36. (a) | 37. (a) | 38. (c) | 39. (a) | 40. (d) | 41. (c) | 42. (a) |
| 43. (a) | 44. (b) | 45. (b) | 46. (c) | 47. (c) | 48. (c) | 49. (a) |
| 50. (c) | 51. (c) | 52. (d) | 53. (d) | 54. (b) | 55. (a) | 56. (b) |
| 57. (a) | 58. (c) | 59. (b) | 60. (c) | 61. (c) | 62. (b) | 63. (b) |
| 64. (b) | 65. (b) | 66. (c) | 67. (b) | 68. (d) | 69. (d) | 70. (a) |
| 71. (c) | 72. (c) | 73. (a) | 74. (d) | 75. (a) | 76. (c) | 77. (c) |
| 78. (a) | 79. (b) | 80. (d) | 81. (a) | 82. (a) | 83. (d) | 84. (c) |
| 85. (b) | 86. (b) | 87. (a) | 88. (a) | 89. (a) | 90. (c) | 91. (b) |
| 92. (b) | 93. (d) | 94. (b) | 95. (b) | 96. (d) | 97. (a) | 98. (b) |
| 99. (c) | 100. (c) | 101. (a) | 102. (a) | 103. (c) | 104. (d) | 105. (b) |
| 106. (a) | 107. (b) | 108. (a) | 109. (c) | 110. (b) | 111. (a) | 112. (c) |
| 113. (d) | 114. (a) | 115. (c) | 116. (a) | 117. (a) | 118. (c) | 119. (b) |
| 120. (c) | | | | | | |

# 9

# ICAR – JRF Horticulture Exam – 2023

1. Given below are two statements, one is labelled as Assertion (A) and other one labelled as Reason (R).

   Assertion (A) : Ditch system of planting is common planting system in Diara land cultivation of cucurbitaceous crops.

   Reason (R) : To manage the availability of moisture and higher temperature, the trenches are dug in North-West direction.

   In light of the above statements, choose the most appropriate answer from the options given below:

   a) Both (A) and (R) are correct and (R) is the correct explanation of (A)

   b) Both (A) and (R) are correct but (R) is NOT the correct explanation of (A)

   c) (A) is correct but (R) is not correct

   d) (A) is not correct but (R) is correct

2. Which of the following is a major pest of the early summer cauliflower?

   a) Diamond back moth b) *Spodoptera litura*

   c) Cabbage butterfly d) Aphid

3. Which of the following chrysanthemum cultivars were developed through induced mutation?

   A. Birbai Sahni B. Pusa Centenary

   C. Pusa Kesari D. Kundan

   Choose the *correct* answer from the options given below:

   a) (A) and (B) only b) (B) and (C) only

   c) (A) and (C) only d) (A) and (D) only

4. Which Irrigation system is useful for areas having water scarcity and salt problems ?

   a) Open cannel b) Sprinkler

   c) Micro sprinkler d) Drip application

5. Match List-I with List-II

| List I | List II |
|---|---|
| Vegetable Crop | Variety/Hybrid |
| A. Tomato | I. Kashi Arun |
| B. French bean | II. Kashi Amul |
| C. Carrot | III. Kashi Himani |
| D. Brinjal | IV. Kashi Baingani |

Choose the *correct* answer from the options given below:

a) (A) - (I), (B) - (II), (C) - (III), (D) - (IV)

b) (A) - (II), (B) - (IV), (C) - (I), (D) - (III)

c) (A) - (III), (B) - (I), (C) - (II), (D) - (IV)

d) (A) - (III), (B) - (IV), (C) - (I), (D) - (II)

6. Given below are two statements:

**Statement (I):** Dwarfing interstocks in apple are used to avoid the undesirable root system of a dwarfing root stock

**Statement (II):** *Mallus Sikimmensis* is largely used as interstock for the said purpose.

In light of the above statements, choose the *most appropriate* answer from the options given below.

a) Both Statement (I) and Statement (II) are true.

b) Both Statement (I) and Statement (II) are false.

c) Statement (I) is true but Statement (II) is false.

d) Statement (I) is false but Statement (II) is true.

7. Which of the following annuals are directly sown in the beds?

A. Pansy B. Nasturtium

C. Stock D. Hollyhock

Choose the *correct* answer from the options given below;

a) (A) and (B) only b) (A) and (C) only

c) (A) and (D) only d) (B) and (D) only

8. Make the correct sequence of steps involved in the T- budding method of rose flower crop?

A. Incision of T shaped cut on the rootstock stem.

B. Selection and planting of rootstocks.

C. Insertion of scion wood in the T-cut on rootstock stern.

D. Selection and preparation of scion wood.

E. Wrapping of scion budded on rootstock stem with polythene.

Choose the *corrcct* answer from the options given below:

a) (A), (B), (D), (E), (C) b) (A), (C), (B), (E), (D)

c) (B), (D), (A), (C), (E) d) (D), (E), (C), (B), (A)

9. Which of the following grafting can be used in tomato if scion and rootstocks are of different thickness?

a) Cleft grafting b) Splice grafting

c) Tube grafting d) Tongue Approach grafting

10. Match List I with List II

| List I (Garden Name) | | List II (Location/state) | |
|---|---|---|---|
| A. | Lalbagh garden | I. | Delhi |
| B. | Amrit Udyan | II. | Tamil Nadu |
| C. | Sim's Park | III. | West Bengal |
| D. | Llyod Botanical Garden | IV. | Karnataka |

Choose the *correct* answer from the options given below:

a) (A) - (III), (B) - (IV), (C) - (II), (D) - (I)

b) (A) - (I), (B) - (II), (C) - (IV), (D) - (III)

c) (A) - (II), (B) - (III), (C) - (IV), (D) - (I)

d) (A) - (IV), (B) -(I), (C) - (II), (D) - (III)

11. Plants which possess special structures such as thorns, rootlets, tendrils, etc to climb over a support, are called as.

a) Seasonal flowers b) Trees

c) Shrubs d) Climbers

12. Given below are two statements, one is labelled as Assertion (A) and other one labelled as Reason (R).

Assertion (A): Potassium accelerates the translocation of photosynthates from leave to tubers, increase higher root yield in sweet potato.

Reason (R): The quality characters like starch and protein contents decreases with increase in potassium in sweet potato.

In light of the above statements, choose the *most appropriate* answer from the options given below:

a) Both (A) and (R) are correct and (R) is the correct explanation of (A).

b) Both (A) and (R) are correct but (R) is NOT the correct explanation of (A).

c) (A) is correct but (R) is not correct.

d) (A) is not correct but (R) is correct.

13. The wheat grain can be stored well at a moisture percentage less than

a) 20% b) 15%

c) 10% d) 5%

14. Exposing seed to beneficial microorganisms is called as

a) Matrix priming b) Hydropriming

c) Haloprirning d) Bio priming

15. Powdery mildew resistant single dominant gene (Er-3) derived from which pea species?

a) *Pisum abyssinicum* b) *Pisum fulvum*

c) *Pisum elatius* d) *Pisum sativum*

16. The world's first cytoplasmic male sterility (CMS) based pigeonpea hybrid is :

a) ICPH2671/Pushkal b) Pargati

c) ICPL85063/Laxmi d) Durga/ICPL84031

17. Given below are two statements, one is lablelled as Assertion (A) and other is labelled as Reason (R)

Assertion (A) : 2, 3, 5 triphenyl tetrazolium chloride (TTC) is used for determining the seed viability

Reason(R) : Living tissue changes the TTC to an insoluble red compound formazan

a) Both (A) and (R) are correct and (R) is the correct explanation of (A).

b) Both (A) and (R) are correct but (R) is not the correct explanation of (A).

c) (A) is correct but (R) is not correct

d) (A) is not correct but (R) is correct

18. Which of the following is NOT a triploid hybrid or triploid variety of Tapioca?

a) Sree Sahya b) Sree Athulya

c) Sree Apoorva d) Sree Harsha

19. Match List-I with List-II

| List-I Variety | List-II Fruit |
|---|---|
| A. Arka Majestic | I. Mango |
| B. PhuleArakta | II. Guava |
| C. ArkaAmulya | III. Pomegranate |
| D. Manjeera | IV. Grapes |

Choose the *correct* answer from the options given below:

a) (A) - (IV), (B) - (III), (C) - (II), (D) - (I)
b) (A) - (I), (B) - (II), (C) - (III), (D) - (IV)
c) (A) - (I), (B) - (II), (C) - (IV), (D) - (III)
d) (A) - (III), (B) - (IV), (C) - (I), (D) - (II)

20. Which of the following are the elements of garden and landscape designs?

a) Line
b) Balance
c) Rhythm
d) Style

Choose the *correct* answer from the options given below:

a) (A) and (C) only
b) (A) and (B) only
c) (B) and (C) only
d) (A) and (D) only

21. Which of the following tree is associated with the life of Lord 'Krishna'?

a) *Salix babylonica*
b) *Anthocephalus cadamba*
c) *Saraca indica*
d) Michelia champaca

22. Given below are two statements:

Statement (I): Worldwide tomato is mainly consumed as fresh vegetable (89%) and only ~11% is processed.

Statement (II) : Tomato genotypes suitable for processing require key parameters such as low TSS (<5 °Brix) and minimum acid: sugar ratio (15:1)

In light of the above statements, choose the *most appropriate* answer from the options given below.

a) Both Statement (I) and Statement (II) are true.
b) Both Statement (I) and Statement (H) are false.
c) Statement (I) is true but Statement (II) is false.
d) Statement (I) is false but Statement (II) is true.

23. When the soil cover of the crop is about 70-80%, the crop coefficient is referred to as

a) Initial stage
b) Mid season stage
c) Crop development stage
d) Late season stage

24. Given below are two statements:

Statement (I) : Each Parasitoids requires only one host, which it kills for its development into a free living adult and are of same size as the host

Statement (II) : Where as predators are free living and often lager in size than their pray requiring several prays to complete their life cycle.

In light of the above statements, choose the most appropriate answer from the options given below.

a) Both Statement (I) and Statement (II) are true
b) Both Statement (I) and Statement (II) are false
c) Statement (I) is true but Statement (II) is false
d) Statement (I) is false but Statement (II) is true

25. Which of the following is not a suitable breeding objective in cowpea?

a) Determinate plant type
b) Early pod maturity
c) Photosensitivity
d) Wider adaptability

26. In guava red spots on the newly emerged leaves and dry and brittle leaves indicates the deficiency of

a) Boron
b) Zinc
c) Phosphorus
d) Potassium

27. Which of the following prostrate or trailing shrubs are suitable for planting in rock gardens?

A. *Lantana sellowiana*
B. *Ixora armeniaca*
C. *Juniperus horizontalis*
D. *Duranta plumeri*

Choose the *correct* answer from the options given below:

a) (A) and (B) only
b) (A) and (C) only
c) (B) and (D) only
d) (A) and (D) only

28. Match List-I with List-II

| List-I Crop | List-II Variety |
|---|---|
| A. Turmeric | I. Pant C-1 |
| B. Coriander | II. Pant Ragini |
| C. Chilli | III. Pant Peetabh |
| D. Fenugreek | IV. Pant Haritima |

Choose the *correct* answer from the options given below:

a) (A) - (III), (B) - (I), (C) - (IV), (D) - (II)

b) (A) - (I), (B) - (II), (C) - (III), (D) - (IV)

c) (A) - (III), (B) - (II), (C) - (I), (D) - (IV)

d) (A) - (III), (B) - (IV), (C) - (I), (D) - (II)

29. Given below are two statements:

Statement (I): Self-incompatibility was first reported by Koelreuter and in case of self- incompatibility, pollen grains fail to germinate on the stigma of flower that produced them.

Statement (II) : In heterornorphic system of self-in compatibility, flowers of different incompatibility groups are different in morphology.

In light of the above statements, choose the *most appropriate* answer from the options given below.

a) Both Statement (I) and Statement (II) are correct.

b) Both Statement (I) and Statement (II) are incorrect.

c) Statement (I) is correct but Statement (II) is incorrect.

d) Statement (I) is incorrect but Statement (II) is correct.

30. Given below are two statements:

Statement (I): In India ratoon crop of sugarcane occupies about 50-55% of total cane area but contributes to only 30-35% of the total cane production

Statement (II) : The varieties possessing high regeneration capacity are required for ratooning and optimum temperature for quick sprouting of stubbles is around 27°C

In light of the above statements, choose the *most appropriate* answer from the options given below.

a) Both Statement (I) and Statement (II) are true.

b) Both Statement (I) and Statement (II) are false.

c) Statement (I) is true but Statement (II) is false.

d) Statement (I) is false but Statement (II) is true.

31. Match List-I with List-II

| List-I Site of Synthesis | List-II Phytohormone |
|---|---|
| A. Young leaves and buds | I. Ethylene |
| B. Roots | II. Cytokinin |
| C. Young fruits | III. GA |
| D. Old leaves | IV. IAA |

Choose the *correct* answer from the options given below:

a) (A) - (I), (B) - (II), (C) - (III), (D) - (IV)

b) (A) - (IV), (B) - (III), (C) - (II), (D) - (I)

c) (A) - (II), (B) - (III), (C) - (IV), (D) - (I)

d) (A) - (II), (B) - (IV), (C) - (I), (D) - (III)

32. Reducing weed growth by manipulating water level in water bodies, light intensities, nutrient availability, competitive displacement, etc. is known as

Preventive weed management

a) Mechanical weed management b) Ecological weed management

c) Cultural weed management

33. Make the correct sequence of steps involved in hybridization of flower crops?

A. Pollination

B. Emasculation of parents

C. Choice and evaluation of parents

D. Harvesting and storing of FI seeds

E. Bagging and tagging

Choose the *correct* answer from the options given below:

a) (C), (D), (E), (B), (A) b) (A), (D), (C), (B), (E)

c) (A), (D), (E), (C), (B) d) (C), (B), (E), (A), (D)

34. Kashi Chayan is the variety of

a) Cowpea b) French bean

c) Tomato d) Chilli

35. Which country proposed UN for declaring the year 2023 as International year of millets?

a) Ethiopia b) China

c) India d) Nigeria

36. Polyembrony is controlled by?

a) Dominant gene b) Recessive gene

c) Single pair of gene d) Self sterility

37. Given below are two statements:

Statement (I): Self pollination leads to a very rapid increase in homozygosity and self pollinated crops do not show inbreeding depression.

Statement (II) : Cross pollination preserves and promotes heterozygosity in a population and shows mild to severe inbreeding depression.

In light of the above statements, choose the ***most appropriate*** answer from the options **given** below.

a) Both Statement (I) and Statement (II) are correct.

b) Both Statement (I) and Statement (II) are incorrect.

c) Statement (I) is correct but Statement (II) is incorrect.

d) Statement (I) is incorrect but Statement (II) is correct.

38. Match List-I with List-II

| List-I Disease | List-II Indicator plant |
|---|---|
| A. Tristeza | I. Kagzilime |
| B. Exocortis | II. Rangpur lime |
| C. Psorosis | III. Sweet orange var. Pineapple |
| D) Greening | IV. Sweet orange var. Valencia |

Choose the *correct* answer from the options given below:

a) (A) - (I), (B) - (II), (C) - (III), (D) - (IV)

b) (A) - (IV), (B) - (II), (C) - (III), (D) - (I)

c) (A) - (I), (B) - (IV), (C) - (II), (D) - (III)

d) (A) - (I), (B) - (III), (C) - (II), (D) - (IV)

39. Banana are placed in the order

a) Rhamnales b) Zinger bales

c) Myrtales d) Resales

40. Which of the following colours belong to soft cool colours in "Colour scheme"?

a) Red, orange and yellow b) White, black and grey

c) Blue, violet and green d) Red, white and green

41. Which of the following physiological disorder exhibit premature initiation of floral buds and elongation of peduncle in cauliflower ?

a) Ricyness b) Fuzziness

c) Leafmess d) Whiptail

42. What is the centre of origin of Fenugreek ?

a) South East Europe and West Asia

b) China

c) Indo-Myanrnar region

d) North and South America

43. Given below are two statements:

Statement (I): Lettuce is most common among leafy vegetables that can be easily grown in hydroponic system.

Statement (II) : Life cycle of hydroponic lettuce is long as compared to traditionally grown lettuce.

In light of the above statements, choose the *most appropriate* answer from the options given below.

a) Both Statement (I) and Statement (II) are true.

b) Both Statement (I) and Statement (II) are false.

c) Statement (I) is true but Statement (II) is false.

d) Statement (I) is false but Statement (II) is true.

44. What is the family of Celery ?

a) Apiaceae  b) Asteraceae

c) Poaceae  d) Amaryllidaceae

45. Papaya hybrid free from typical papaya odour

A. ArkaPrabhath  B. Arka Surya

C. CO 5  D. Solo

a) (A), (B) and (c) only  b) (A) and (B) only

c) (A), (B) and (D)  d) (B), (C) and (D) only

46. What is the spacing (row to row & plant to plant) of extra early varieties of cauliflower?

a) 45 cm x 30 cm  b) 60 cm x 45 cm

c) 60 cm x 30 cm  d) 75 cm x 60 cm

47. Which of the following are cool season turf grasses?

A. *Cynodon dactylon*  B. *Agrostis palustris*

C. *Paspalum notation*  D. *Lolium perene*

Choose the *correct* answer from the options given below;

a) (A) and (B) only  b) (A) and (C) only

c) (B) and (D) only  d) (C) and (D) only

48. Which of the following is the most common rootstock for budding in rose under South Indian conditions?

a) *Rosa foetida*  b) *Rosa multiflora*

c) *Rosa lucida*  d) Rosa indicavar.odorata

49. Sequence the Cross combinations of the apple rootstock

A. Mailing 2 x Northern Spy B. Mallingl x Northern Spy

C. Northern Spy x Merton 793 D. Northern Spy x Mailing 2

Choose the *correct* answer from the options given below;

a) M.25-M.111-M.104-M.106 b) M.111-M.25-M.106-M.104

c) M.104-M.106-M.111-M.25 d) M.104-M.111-M.106-M.25

50. Self unfruitful cultivar of peach

a) J.H. Hale b) Maygold

c) Sharbati d) Candor

51. Match List-I with List-II

| List-I Characteristic | | List-II Fruit | |
|---|---|---|---|
| A. | Fleshy peduncle | I. | Cashew |
| B. | Non climacteric | II. | Pomegranate |
| C. | Drupe | III. | Mango |
| D. | Synconium | IV. | Fig |

Choose the *correct* answer from the options given below:

a) (A) - (I), (B) - (II), (C) - (III), (D) - (IV)

b) (A) - (II), (B) - (I), (C) - (III), (D) - (IV)

c) (A) - (I), (B) - (II), (C) - (IV), (D) - (III)

d) (A) - (III), (B) - (IV), (C) - (I), (D) - (II)

52. Given below are two statements, one is lablelled as Assertion (A) and other is labelled as Reason (R)

Assertion (A): Gerrnplasrn can be preserved for longer period under cryopreservation at -296°C.

Reason (R): There is minimal loss of genetic information through cryopreservation

In light of above statements, choose the *most appropriate* answers from the options given below.

a) Both (A) and (R) are correct and (R) is the correct explanation of (A).

b) Both (A) and (R) are correct but (R) is not the correct explanation of (A).

c) (A) is correct but (R) is not correct

d) (A) is not correct but (R) is correct

53. Variety 'Pusa Sharda' belongs to which crop that can be stored for 15-20 days under room temperature ?

a) Tomato b) Cucumber

c) Muskmelon d) Bitter gourd

54. Respiration

A. Occurs only in cells with chloroplasts

B. Oxygen is used and water is produced

C. Occurs in dark and light

D. Occurs in light only

Choose the *correct* answer from the options given below:

a) (A), (B) and (D) only b) (A) and (C) only

c) (A), (B), (C) and (D) d) (B), (C) and (D) only

55. Given below are two statements, one is labelled as Assertion (A) and other one labelled as Reason (R).

Assertion (A): Most vegetables except onion and garlic are blanched before drying

Reason (R): Blanching stops the action of enzymes, which are not killed by sun drying

In light of the above statements, choose the *correct* answer from the options given below.

a) Both (A) and (R) are true and (R) is the correct explanation of (A).

b) Both (A) and (R) are true but (R) is NOT the correct explanation of (A).

c) (A) is true but (R) is false.

d) (A) is false but (R) is true.

56. Given below are two statements, one is labelled as Assertion (A) and other one labelled as Reason (R).

Assertion (A) : Calyx splitting is a major disorder of carnation and in this, calyx split when the flowers start to open.

Reason (R): This is attributed to various factors such as over feeding, varietal character, high temperature, a sharp drop in night temperature, a low nitrogen level or boron deficiency.

In light of the above statements, choose the *most appropriate* answer from the options given below:

a) Both (A) and (R) are correct and (R) is the correct explanation of (A).

b) Both (A) & (R) are correct but (R) is NOT the correct explanation of (A).

b) (A) is correct but (R) is not correct.

d) (A) is not correct but (R) is correct.

57. Given below are two statements:

Statement (I): Gigas and genome buffering effect exhibited by haploid plants

Statement (II): In watermelon, there is no existence of parthenocarpy under natural conditions.

In light of the above statements, choose the *most appropriate* answer from the options given below.

a) Both Statement (I) and Statement (II) are correct.

b) Both Statement (I) and Statement (II) are incorrect.

c) Statement (I) is correct but Statement (II) is incorrect.

d) Statement (I) is incorrect but Statement (II) is correct.

58. Choose the right sequence (decreasing order) about India's share of the following fruits in global fruit production

A. Mango B. Grapes

C. Banana D. Citrus

Choose the *correct* answer from the options given below:

a) (A), (B), (C), (D) b) (A), (C), (D), (B)

c) (B), (A), (D), (C) d) (C), (B), (D), (A)

59. Given below are two statements, one is lablelled as Assertion (A) and other is labelled as Reason (R)

Assertion (A): Clonal selection followed by vegetative propagation is a major strategy for improvement of perennial crops.

Reason (R): A large genetic advance can be made in a single step without the multiple generation of seed propagation required of seed propagated cultivars.

In light of above statements, choose the *most appropriate* answers from the options given below.

a) Both (A) and (R) are correct and (R) is the correct explanation of (A)

b) Both (A) & (R) are correct but (R) is not the correct explanation of (A)

c) (A) is correct but (R) is not correct

d) (A) is not correct but (R) is correct

60. Which of the following biocontrol agent is recommended to control diamond back moth (DBM), serious pests of cruciferous crops particularly cabbage and cauliflower ?

a) *Brachymeria lasus* a) *Cotesia plutellae*

a) *Trichogramma pretiosum* a) *Chrysoperla zastrowiilletni*

61. Citrus rootstock incompatible with lime and lemon

a) *Citrus unshiu*  b) Rangpur lime

c) Kama Khatta  d) Trifoliate orange

62. Which of the following climber is categorized as heavy climber?

a) *Petrea volubilis*  b) *Quisqualis indica*

c) *Aristohchia elegans*  d) *Lonicera japonica*

63. Match List-I with List-II

| List-I | List-II |
|---|---|
| Propagtion Method | Other name |
| A. Marcottage | I. Air Layering |
| B. Stooling | II. Mound layering |
| C. Serpentine | III. Compound layering |
| D. Inarching | IV. Attached grafting |

Choose the *correct* answer from the options given below:

a) (A) - (I), (B) - (II), (C) - (III), (D) - (IV)

b) (A) - (I), (B) - (III), (C) - (II), (D) - (IV)

c) (A) - (III), (B) - (II), (C) - (I), (D) - (IV)

d) (A) - (III), (B) - (IV), (C) - (I), (D) - (II)

64. Which pineapple cultivar has spineless leaves and is resistant to gurnmosis?

a) Smooth Cayenne  b) Queen

c) Spanish  d) Abacaxi

65. Which of the following is a largest Indian orchid?

a) *Bletia purpurea*  b) *Aerides multiflorum*

c) *Cytnbidium giganteurn*  d) *Galeola falconeri*

66. First hybridization work in citrus was carried out at

a) Kodur  b) Meghalaya

c) Nagpur  d) Chethali

67. Yearly sequence and spatial arrangement of crops and fallow on a given area is

a) Crop rotation  b) Cropping pattern

c) Cropping system  d) Double cropping

68. Given below are two statements:

Statement (I): Mutagen induced male sterility is heritable and male sterility induced by gametocide is non-heritable.

Statement (II): Heterozygous population derived in garden pea will not segregate on selfing.

In light of die above statements, choose the *most appropriate* answer from the options given below.

a) Both Statement (I) and Statement (II) are true.

b) Both Statement (I) and Statement (II) are false.

c) Statement (I) is true but Statement (II) is false.

d) Statement (I) is false but Statement (II) is true.

69. Sequence the mango varieties according to their parentage

A. Amrapali x Sensation | B. Dashehari x Sensation

C. Amrapali x Vanraj | D. Amrapali x Lal Sundari

Choose the *correct* answer from the options given below:

a) Pusa Lalima - Pusa Peetamber - Arunika - Pusa Shreshth

b) Pusa Shreshth Pusa Lalima - Arunika - Pusa Peetamber

c) Pusa Shreshth - Arunika - Pusa Peetamber - Pusa Lalima

d) Pusa Shreshth - Pusa Lalima - Arunika - Pusa Peetamber

70. Life cycle of a normal cultivated rice from germination to physiological maturity is completed in

a) 80-90 Days

b) 90-100 days

c) 110-210 Days

d) 250-260 Days

71. Match List-I with List-II

| List-I<br>Types of hydroponics | List-II<br>Functional Mechanism |
|---|---|
| A. Wick system | I. Plants stand in a shallow stream of water containing desirable nutrients |
| B. Ebb and Flow | II. Net pot filled in a small quantity of clay pebbles |
| C. NFT system | III. Passive form of hydroponics |
| D. Water culture | IV. Periodic flooding and draining of the nutrient solutions |

Choose the *correct* answer from the options given below:

a) (A) - (II), (B) - (IV), (C) - (I), (D) - (III)

b) (A) - (III), (B) - (I), (C) - (IV), (D) - (II)

c) (A) - (IV), (B) - (III), (C) - (I), (D) - (II)

d) (A) - (III), (B) - (IV), (C) - (I), (D) - (II)

72. Given below are two statements, one is labelled as Assertion (A) and other one labelled as Reason (R).

Assertion (A): Shrubs or trees planted at an irregular interval to form a discontinuous screen is known as hedge.

Reason (R): An ideal hedge is having uniform dense growth from base to top and it can be achieved by getting more or less uniform supply of water, essential elements and proper sunshine.

In light of the above statements, choose the most appropriate answer from the options given below:

a) Both (A) and (R) are correct and (R) is the correct explanation of (A).

b) Both (A) & (R) are correct but (R) is NOT the correct explanation of (A).

c) (A) is correct but (R) is not correct.

d) (A) is not correct but (R) is correct.

73. The calcium content in healthy plant ranges between

a) 0.05-5.0%  b) 0.1-0.3.0%

c) 0.2-1.0%  d) 0.5-5.0%

74. The recommended optimum plant density for sweet potato has been recommended by ICAR- CTCRI is ?

a) 73000 plants/ha  b) 63000 plants/ha

c) 83000 plants/ha  d) 93000 plants/ha

75. Match List-I with LIst-II

| List-I | | List-II | |
|---|---|---|---|
| Botanical Name | | Vegetable Crop | |
| A. | *Betiincasa hispida* | I. | Lotus |
| B. | *Trapa natans* | B. | Winged bean |
| C. | *Nelumbo nucifera* | III. | Ash gourd |
| D. | *Psophocarpus tetragonolobus* | IV. | Water chestnut |

Choose the *correct* answer from the options given below:

1. (A) - (I), (B) - (II), (C) - (III), (D) - (IV)
2. (A) - (III), (B) - (II), (C) - (I), (D) - (IV)
3. (A) - (I), (B) - (II), (C) - (IV), (D) - (III)
4. (A) - (III), (B) - (IV), (C) - (I), (D) - (II)

76. Match List-I with List-II

| List-I | List-II |
|---|---|
| Physiological disorder | Fruit crop |
| A. Translucent flesh | I. Mango |
| B. Internal breakdown | II. Pineapple |
| C. Gooseflesh | III. Pomegranate |
| D. Girdle necrosis | IV. Banana |

Choose the *correct* answer from the options given below:

1. (A) - (I), (B) - (II), (C) - (III), (D) - (IV)
2. (A) - (II), (B) - (III), (C) - (IV), (D) - (I)
3. (A) - (I), (B) - (II), (C) - (IV), (D) - (III)
4. (A) - (III), (B) - (IV), (C) - (I), (D) - (II)

77. Which Apple cultivar is a good pollinizer for all the delicious groups ?

a) Jonathan b) Jona Gold

c) Golden Delicious d) Greensleeves

78. Asepsis is

a) Completely destroying the microorganism by high temperature

b) Keeping microorganism out from entering into food

c) Removal of microorganism from food by filtering

d) It is a type of freezing

79. Match List I with List II according to the synonym used for the species

| List I | List II |
|---|---|
| A. *Vitis vinifera* | I. Cat grape |
| B. *Vitis labrusca* | II. Sand grape |
| C. *Vitis tupestris* | III. Fox grape |
| D. *Vitis palmata* | IV. Wine grape |

Choose the *correct* answer from the options given below:

a) (A) - (I), (B) - (III), (C) - (IV), (D) - (II)

b) (A) - (III), (B) - (IV), (C) - (I), (D) - (II)

c) (A) - (IV), (B) - (III), (C) - (II), (D) - (I)

d) (A) - (II), (B) - (I), (C) - (III), (D) - (IV)

80. Which of the following is a striped variety of standard carnation?

a) Darling b) Picaro

c) Super Star d) Challenger

81. Given below are two statements, one is labelled as Assertion (A) and other one labelled as Reason (R).

Assertion (A): Making a notch below a bud by removing a wedge shaped piece of bark is known as nicking and it helps in formation of fruit bud

Reason (R): Nicking ensures accumulation of carbohydrates from the leaves to the bud which help in the formation of fruit bud

In light of the above statements, choose the *correct* answer from the options given below.

a) Both (A) and (R) are true and (R) is the correct explanation of (A).

b) Both (A) and (R) are true but (R) is NOT the correct explanation of (A).

c) (A) is true but (R) is false.

d) (A) is false but (R) is true.

82. What is the new name of'Mughal Garden' of Rashtrapati Bhavan, New Delhi, India?

a) Amrit Vatika b) Parmo Udyan

c) Amrit Udyan d) Nandan Van

83. Which mango cultivar imparts prominent beak character to the progenies?

a) Manjeera b) Totapuri

c) Banganpalli d) Amrapalli

84. Which of the following garden styles are developed on the basis of their respective ideas of heaven?

A. Mughal garden style B. Persian Garden style

C. English Garden style D. Japanese garden style

Choose the *correct* answer from the options given below:

a) (A) and (C) only b) (A) and (D) only

c) (B) and (D) only d) (C) and (D) only

85. Match List-I with List-II

| List-I (Flower Crop) | List-II (Basic chromosome number) |
|---|---|
| A. Rose | I. X = 9 |
| B. Gladiolus | II. X = 12 |
| C. Marigold | III. x = i5 |
| D. Chrysanthemum | IV. X = 7 |

Choose the *correct* answer from the options given below:

1. (A) - (I), (B) - (II), (C) - (IV), (D) - (III)
2. (A) - (III), (B) - (IV), (C) - (II), (D) - (I)
3. (A) - (II), (B) - (III), (C) - (IV), (D) - (I)
4. (A) - (IV), (B) - (III), (C) - (II), (D) - (I)

86. What is the causal organism of early blight of potato ?

a) Alternaria solani
b) Phytophthora infestans
c) Rhizoctonia solani
d) Synchytrium endobioticum

87. Given below are two statements:

Statement (I): The tall coconut cultivars are referred as var. typica, while dwarf type as var. Nana

Statement (II): Tall cultivars are the resultant of out breeding while dwarf are the resultant of inbreeding

In light of die above statements, choose the *most appropriate* answer from the options given below.

a) Both Statement (I) and Statement (II) are correct.
b) Both Statement (I) and Statement (II) are incorrect.
c) Statement (I) is correct but Statement (II) is incorrect.
d) Statement (I) is incorrect but Statement (II) is correct.

88. Which of the following is FI hybrid of watermelon ?

a) Arka Akash
b) Arka Shyania
c) Arka Manik
d) Arka Muthu

89. Statistical analysis for determining the adaptability of genotypes in different environment is

a) Correlation analysis
b) Line x Tester
c) $D^2$ Mahalanobis
d) Stability analysis

90. Given below are two statements, one is labelled as Assertion (A) and other one labelled as Reason (R).

Assertion (A): Establishment of cashew orchards by seedling is not recommended

Reason (R): Cashew is a cross- pollinated crop

In light of the above statements, choose the *correct* answer from the options given below.

a) Both (A) and (R) are true and (R) is the correct explanation of (A).

b) Both (A) and (R) are true but (R) is NOT the correct explanation of (A).

c) (A) is true but (R) is false.

d) (A) is false but (R) is true.

91. Given below are the two statements:

Statement (I): Ambred is cross between Red Delicious X Amberi

Statement (II): The Ambred tree is very dwarf

Based on these statement choose the correct option:

a) Both Statement (I) and Statement (II) are true.

b) Both Statement (I) and Statement (II) are false.

c) Statement (I) is true but Statement (II) is false.

d) Statement (I) is false but Statement (II) is tine.

92. Famous blue coloured variety 'Moondust' belongs to which of the following flower crop?

a) Lilium longiflorutn   b) Rosa hybrida

c) Dianthus caryophyllus   d) Tagetes erecta

93. Match List I with List 11 pear rootstock for their salient characteristics

| List I | List II |
|---|---|
| A. *Pyrus betulifolia* | I. Blight resistant rootstock |
| B. *Cyndonia oblonga* | II. Black end susceptible rootstock |
| C. *Pyrus calleryana* | III. Dwarfing rootstock |
| D. *Pyrus pyrifolia* | IV. Alkaline tolerant rootstock |

Choose the *correct* answer from the options given below:

a) (A) - (I), (B) - (II), (C) - (III), (D) - (IV)

b) (A) - (III), (B) - (IV), (C) - (II), (D) - (I)

c) (A) - (IV), (B) - (III), (C) - (I), (D) - (II)

d) (A) - (II), (B) - (I), (C) - (IV), (D) - (III)

94. Which Mallus species is known as Siberian crab?

a) *M. baccata* b) *M. coronaria*

c) *M. floribunda* d) *M. angustifolia*

95. Given below are two statements, one is lablelled as Assertion (A) and other is labelled as Reason (R).

Assertion (A): Nucellar embryony has been a major obstacle in citrus breeding

Reason (R): Nucellar embryos arise from somatic cells of the nucellus,they are asexual progenies and not required by the breeder.

In light of above statements, choose the *most appropriate* answers from the options given below.

a) Both (A) and (R) are correct and (R) is the correct explanation of (A)

b) Both (A) & (R) are correct but (R) is not the correct explanation of (A)

c) (A) is correct but (R) is not correct

d) (A) is not correct but (R) is correct

96. Sequence the fruit crop according to the place of their origin

A. Indo Burma region B. South East Asia

C. Black Sea - Caspian sea D. Mesoamerica

Choose the *correct* answer from the options given below:

a) Grape - Mango - Papaya - Banana

b) Banana - Mango - Papaya - Grape

c) Mango — Banana — Grape - Papaya

d) Papaya - Grape - Mango - Banana

97. On the basis of anthesis time of vegetable crops

A. Tomato B. Muskinelon

C. Summer Squash D. Ridge gourd

Choose the *correct* answer from the options according to ascending order of athesis time:

a) (B), (C), (D), (A) b) (A), (B), (C), (D)

c) (B), (A), (D), (C) d) (C), (B), (A), (D)

98. *Dicladispa annegera* is major pest of

a) Rice a) Wheat

a) Sugarcane a) Gram

99. A. KalpaSankara is coconut hybrid with cross between CGD and WCT

    B. KalpaSankara is coconut hybrid with cross between MYD and WCT

    C. High yelding

    D. Tolerant to root (wilt) disease

    Choose the *correct* answer from the options given below:

    a) (A), (C) and (D) only    b) (A), (B) and (D) only

    c) (A), (B), (C) and (D)    d) (B), (C) and (D) only

100. The origin of wound induced De Novo adventitious root in citrus is

    a) Cambium    b) Phloem

    c) Bark and Basal callus    d) Nodal Segment

101. Match List I with List II with respect to salient characteristics of mango cultivar

| List I | List II |
|---|---|
| A. Suvarnarekha | I. Regular bearing |
| B. Rumani | II. Imparting tolerance to Bacterial Canker |
| C. Banganpalli | III. Good donor for skin colour |
| D. Bombay green | IV. Good source for dwarfness |

Choose the *correct* answer from the options given below:

a) (A) - (I), (B) - (III), (C) - (IV), (D) - (II)

b) (A) - (III), (B) - (IV), (C) - (I), (D) - (II)

c) (A) - (IV), (B) - (II), (C) - (III), (D) - (I)

d) (A) - (II), (B) - (I), (C) - (IV), (D) - (III)

102. Given below are two statements:

Statement (I): ArkaMridula is a white pulp guava variety

Statement (II): It is seedling selection from Allahabad Safeda

In light of the above statements, choose the most appropriate answer from the options given below.

a) Both Statement (I) and Statement (II) are true.

b) Both Statement (I) and Statement (II) are false.

c) Statement (I) is true but Statement (II) is false.

d) Statement (I) is false but Statement (II) is true.

103. PramparagatKrishiVikasYojana (PKVY) was launched during the year?

a) 2000 b) 2005

c) 2010 d) 2015

104. What is the basic chromosome number is present in C genome of *Brassica oleracea* ?

a) X = 18 b) x = 9

c) x = 36 d) x = 16

105. Which of the following onion variety is recommended *foikharif* season ?

a) Bhima Light Red b) Bhima Kiran

c) Bhima Dark Red d) Pusa Madhavi

106. Carbohydrates are

A. Molecules that contain carbon, hydrogen and oxygen atoms in specific ratios.

B. Sparing protein

C. Produce energy

D. Prevent lipid metabolism

Choose the *correct* answer from the options given below:

a) (A), (B) and (D) only b) (A), (B) and (C) only

c) (A), (B), (C) and (D) d) (B), (C) and (D) only

107. What is the chromosome number of Radish ?

a) 2n = 16 b) 2n = 18

c) 2n = 20 d) 2n = 22

108. Which Coimbatore series of Papaya is a selection from Washington?

a) Co-1 b) Co-3

c) Co-5 d) Co-7

109. Dwarf palms are

A. Referred as var. nana

B. Short stature, 8-10 M at 20 years of age

C. Start bearing at the age **of** 3-4 years

D. Have long productive life of more than 100 years

E. Chowghat Green Dawrf and Chowghat orange are popular cultivars in India

Choose the *correct* answer from the options given below:

a) (A). (B) and (D) only b) (A), (B), (C) and (E) only

c) (A), (B), (C) and (D) only d) (B), (C), (D) and (E) only

110. Improper filling physiological disorder of banana is due to the deficiency of

a) Iron b) Potassium

c) Copper d) Magnesium

111. Micronutrient that detoxifies super oxide free radicals, proteins and lipids against oxidation is

a) Zinc b) Boron

c) Molybdenum d) Chlorine

112. Which of the following shrubs are grown for their fragrant flowers?

A. *Duranta plwnieri* B. *Cestrum nocturnum*

C. *Acalypha hispida* D. *Portlandia grandiflora*

Choose the *correct* answer from the options given below;

a) (A) and (C) only b) (B) and (C) only

c) (B) and (D) only d) (C) and (D) only

113. The grape vine flowers are borne in clusters or panicles on

a) 1-2 node b) 3-5 node

c) 5-7 node d) 8-10 node

114. Which of the following have most desirable traits for edible pod ?

a) Parchment free, thin-walled pods

b) Semi-parchment, thick-walled pods

c) Parchment free, thick-walled pods

d) Thin walled pods with parchment layer

115. Which of the following variety is suitable for early sowing in garden pea ?

a) Pusa Shree b) Pusa Prabal

c) PalamPriya d) ArkaAjil

116. Calculate the number of plants in qunicunx system of planting in one hectare area when row to row and plant to plant distance is 10 m

a) 90 b) 150

c) 181 d) 200

117. Which of the following annual flower crop is used as green filler?

a) Nasturtium b) Rose

c) Gypsophila d) Carnation

118. A dried drupe of black pepper is known as

a) Pepper corn
b) Pepper berry
c) Pepper spikes
d) Pepper leaf

119. In tomato cultivation, which of the following is used for fruit set under adverse condition ?

a) $GA_3$
b) PCPA
c) Kinetin
d) IBA

120. Given below are two statements, one is labelled as Assertion (A) and other one labelled as Reason (R).

Assertion (A): Long day types onions are cultivated in temperate or hilly regions.

Reason (R): The onion crop is more sensitive to temperature that plays more important role in bulb formation than photoperiod.

In light of the above statements, choose the most appropriate answer from the options given below:

a) Both (A) and (R) are correct and (R) is the correct explanation of (A).
b) Both (A) & (R) are correct but (R) is NOT the correct explanation of (A).
c) (A) is correct but (R) is not correct.
d) (A) is not correct but (R) is correct.

## Answer Key

| | | | | | | | | | | | | | |
|---|---|---|---|---|---|---|---|---|---|---|---|---|---|
| 1 | (c) | 2 | (b) | 3 | (a) | 4 | (a) | 5 | (b) | 6 | (a) | 7 | (b) |
| 8 | (b) | 9 | (c) | 10 | (c) | 11 | (d) | 12 | (a) | 13 | (c) | 14 | (c) |
| 15 | (d) | 16 | (a) | 17 | (a) | 18 | (b) | 19 | (c) | 20 | (b) | 21 | (a) |
| 22 | (a) | 23 | (a) | 24 | (a) | 25 | (a) | 26 | (b) | 27 | (d) | 28 | (a) |
| 29 | (a) | 30 | (a) | 31 | (b) | 32 | (a) | 33 | (b) | 34 | (b) | 35 | (c) |
| 36 | (b) | 37 | (b) | 38 | (c) | 39 | (a) | 40 | (c) | 41 | (a) | 42 | (a) |
| 43 | (b) | 44 | (d) | 45 | (b) | 46 | (b) | 47 | (a) | 48 | (c) | 49 | (a) |
| 50 | (a) | 51 | (a) | 52 | (a) | 53 | (d) | 54 | (a) | 55 | (c) | 356 | (c) |
| 57 | (b) | 58 | (c) | 59 | (c) | 60 | (b) | 61 | (b) | 62 | (b) | 63 | (d) |
| 64 | (a) | 65 | (a) | 66 | (c) | 67 | (c) | 68 | (c) | 69 | (c) | 70 | (d) |
| 71 | (c) | 72 | (b) | 73 | (b) | 74 | (c) | 75 | (a) | 76 | (b) | 77 | (c) |
| 78 | (a) | 79 | (a) | 80 | (a) | 81 | (b) | 82 | (c) | 83 | (d) | 84 | (a) |
| 85 | (a) | 86 | (d) | 87 | (c) | 88 | (a) | 89 | (b) | 90 | (a) | 91 | (d) |
| 92 | (a) | 93 | (c) | 94 | (c) | 95 | (d) | 96 | (d) | 97 | (c) | 98 | (b) |
| 99 | (c) | 100 | (b) | 101 | (d) | 102 | (a) | 103 | (d) | 104 | (c) | 105 | (a) |
| 106 | (b) | 107 | (c) | 108 | (c) | 109 | (d) | 110 | (c) | 111 | (b) | 112 | (a) |
| 113 | (b) | 114 | (c) | 115 | (d) | 116 | (c) | 117 | (d) | 118 | (d) | 119 | (c) |
| 120 | (d) | | | | | | | | | | | | |

# 10

# ICAR – SRF Horticulture Exam – 2015

1. The nodal agency for international co-operation in the area of agricultural research and education in India is:

   a) ICAR b) DARE

   c) CAU d) CSIR

2. Which of the following countries is not a member of G-8?

   a) Canada b) france

   c) Germany d) Spain

3. FSSAI does not provide licensing for:

   a) Non-alcoholic beverage b) Alcoholic beverage

   c) Milk products d) Organic products

4. 'Swachh Bharat Mission' was launced on by the Prime minister of India.

   a) 15th August 2014 b) 2nd October 2014

   c) 16th October 2014 d) 14th November 2014

5. The Nobel Peace Prize 2014 was given to Kailash Satyartbi for his contribution to:

   a) Bachpan fiachao Andolan b) Ganga Bachao Andotan

   c) Bcej Bachao Andolan d) Pani Bachao Andotan

6. The first IVF (in-vitro fertilization) buffalo calf of the world is:

   a) Garima b) Mahitna

   c) Pratham d) Pumima

7. Which of the following is used as an indicator for sulphur dioxide (SO) pollution?

   a) Lichen b) Fern

   c) Fungus d) Algae

8. Which of the following is a CAM (Crassnlacean acid metabolism) plant:
   a) Sugareanc b) Papaya
   c) Rice d) Pineapple
9. 'Kisan Credit Card' (KCC), a credit card to provide affordable credit for farmers in India was started by:
   a) SBI b) NABARD
   c) CB1 d) RRB
10. The funding pattern of NREGA is:
   a) 100% by central government b) 50% by central and 50% by state
   c) 75% by central & 25% by stated) 90% by central and 10% by state
11. Most commonly applied law in agriculture is:
   a) Law of supply b) Law of marginal utility
   c) Law of diminishing return d) Law of demand
12. UN General Assembly declared 2014 as:
   a) International Year of Soil
   b) International Year of Pulses
   c) International Year of Sustainable Agriculture
   d) International Year of Family Farming
13. A3P of NFSM is associated with:
   a) Accelerated palm production Programme
   b) Accelerated cereal production Programme
   c) Accelerated pulses production Programme
   d) Accelerated oilseed production Programme
14. WIPO (World Intellectual Property Organization) is located at:
   a) Paris b) Rome
   c) Geneva d) Washington D.C.
15. Which of the following is/are bio fertilizer (s)?
   a) Rhizobium b) PSB
   c) BGA d) All of these
16. The Indian Standards Institution (ISI) is presently known as:
   a) BS1 b) BIS
   c) ISO d) BISO

17. Golden rice is a rich source of:
    a) Vitamin A b) Lysme
    c) Protein d) Vitamin B1
18. White bud disorder of maize occurs due to deficiency of:
    a) Fe b) Mg
    c) B d) Zn
19. If E=5 and BAS1S=50: WORLD=?
    a) 42 b) 62
    c) 72 d) 82
20. In a 100 m race, A covers the distance in 36 seconds and B in 45 seconds. In this race A beat B by:
    a) 20 m b) 25 m
    c) 22.5 m d) 9 m
21. Cashew nut is commercially propagated by:
    a) Wedge grafting b) Rpicotyl grafting
    c) Tongue grafting d) Softwood grafting
22. Planting density of rubber is:
    a) 250-320 plants/ha. b) 320-400 plants/ha.
    c) 420-450 plants/ha. d) 450-500 plants/ha.
23. The term 'Climacteric' is first used by:
    a) Kidd and West b) Gane
    c) Cruess d) Bkekar
24. Pusa seb mulvrint-1 is a selection from:
    a) Malus baccata (Assam) b) Malus baccata (Khrot)
    c) Malus baccata (Rohru) d) Malus baccata (Shilong)
25. Spacing followed in meadow orcharding is
    a) 1x2 m b) 2x5m
    c) 5x 5m d) 2x3m
26. Localized graft incompatibility can be overcome by:
    a) Insertion of mutually compatible inter-stock
    b) Insertion of mutually compatible scion
    c) Rejuvenation of incompatible grafts
    d) All the above

27. Avocado flowers are:
    a) Morphologically bisexual functionally unisexual
    b) Functionally bisexual morphologically unisexual
    c) Morphologically and functionally unisexual
    d) Morphologically and functionally bisexual
28. Which of the following is a transgenic variety of papaya?
    a) Sinta b) Hortus Gold
    c) Red lady d) Rainbow
29. Conservation of germplasm in their natural habitat is known as:
    a) Ex-situ conservation b) In-situ conservation
    c) Natural conservation d) None of the above
30. Daru, a wild type of pomegranate found in Himalayan region is resistant to:
    a) Fruit cracking b) Blight
    c) Leaf spot d) All the above
31. Hand pollination is practiced in:
    a) Persimmon b) Avocado
    c) Loquat d) Custard apple
32. Emergence of new growth flush immediately after harvesting in mango is an indicator of:
    a) Regularity in bearing b) Dwarfness
    c) Precocity d) Alternate bearing
33. Which of the following is/are indicator of dwarfness in mango?
    a) Higher phloem/xylem ratio b) Lower stomatal density
    c) Higher phenolics d) All the above
34. The temperature required for pollen germination in apple is:
    a) 10-15°C b) 15-20° C
    c) 20-25° C d) 25-30° C
35. Red colour of tamarind is due to:
    a) Leucoanthocynin b) Anthocyanin
    c) Leupol d) Lycopene

36. HACCP stands for:
    a) Hazard analysis critical control points
    b) Hazard analysis critical control pigments
    c) Harmful analysis critical control points
    d) Hazard analysis chronological control points
37. A citrus species which originated from Japan is:
    a) Citrus rugulose b) Citrus macrophylla
    c) Citrus unshu d) Citrus auratntium
38. Which of the following contains no roots at nodes?
    a) Runners b) Suckers
    c) Corms d) Stolen
39. Chilling requirement of subtropical peaches is:
    a) 100-125 hrs. b) 250-400 hrs.
    c) 400-700 hrs. d) 700-1000 hrs.
40. The causal organism of peach leaf curl disease is:
    a) Virus b) Bacteria
    c) Fungus d) Mycoplasma
41. The chamber temperature maintained in vapour heat treatment of mango is 50-52°C while the temperature of mango pulp is:
    a) 45°C b) 46°C
    c) 47°C d) 48°C
42. Long term storage of lemon at low temperature (>13°C) causes chilling injury, a physiological disorder in which internal browing develops in the membrane of the segments and core is known as:
    a) Olecoellosis b) Mambranosis
    c) Granulation d) Rind pitting
43. The phenomenon in which different parts of a plant show phase variation and if propagated through meristems, the different phases are perpetuated in the offspring. This is known as:
    a) Chimera b) Topophysis
    c) Periphysis d) Mutation
44. Optimum condition for storage of mango is:
    a) 12-14°C Temperature and 60-70% RH
    b) 12-14°C Temperature and 80-90% RH
    c) 14-16°C Temperature and 60-70% RH
    d) 14-16°C Temperature and 80-90% RH

45. Which of the following countries has the highest area under protected cultivation?
    a) China b) Japan
    c) USA d) Brazil
46. The ratio of coco peat, vermiculite and perlite used in nursery for rooting medium is:
    a) 1:1:1: b) 2:1:1
    c) 3:1:1 d) 3:2:1
47. Dwarf Cavendish banana takes ......... months to harvest after planting?
    a) 4 to 6 months b) 6 to 10 months
    c) 10 to 14 months d) 14 to 16 months
48. The quality of papain (papaya latex) is measured by:
    a) Methionine b) Tyrosine
    c) Lysine d) None of these
49. Black heart in apricot is a:
    a) Physiological disorder b) Nutritional disorder
    c) Fungal disease d) Bacterial disease
50. Fluctuation in temperature in mango causes:
    a) Vegetative growth b) Flower drop
    c) Malformation d) All of these
51. Highest oil palm producing country in the world is:
    a) Malaysia b) Indonesia
    c) Thailand d) Colombia
52. Which of the following is a leading state in coffee production in India?
    a) Kerala b) Karnataka
    c) Tamil Nadu d) Andhra Pradesh
53. Which of the following varieties of tea is resistant to drought and winds and suitable for higher elevation?
    a) UPASI-1 b) UPASI-7
    c) UPASI-9 d) UPASI-10
54. Which of the following is a non-climacteric fruit?
    a) Peach b) Avocado
    c) Carambola d) Pear

55. Which of the following is a co-dominant marker?

| a) | RAPD | b) | SSR |
|---|---|---|---|
| c) | ISSR | d) | AFLP |

56. Which of the following molecular markers is used for identification of clones?

| a) | RAPD | b) | ISSR |
|---|---|---|---|
| c) | SNP | d) | SSR |

57. Which of the following is a spontaneous hybrid of coffee?

| a) | San Ramon | b) | Cauvery |
|---|---|---|---|
| c) | Hybrido-De-Timor | d) | Old Chicks |

58. The term 'Boroja/Bareja' is related with:

| a) | Black pepper | b) | Betel nut |
|---|---|---|---|
| c) | Betel vine | d) | Cocoa |

59. Sexual dimorphism in betel vine is associated with:

| a) | Leaf size and shape | b) | Stem length and width |
|---|---|---|---|
| c) | Leaf length and width | d) | Leaf length and texture |

60. Light intensity is measured by:

| a) | Light meter | b) | Lux meter |
|---|---|---|---|
| c) | Pyranometer | d) | Psychrometer |

61. CAD stands for:

| a) | Commonly - aided design | b) | Computer - aided design |
|---|---|---|---|
| c) | Computerized - aided design | d) | Company - aided design |

62. Directorate of Floricultural Research (DFR) is located at:

| a) | New Delhi | b) | Bangalore |
|---|---|---|---|
| c) | Pune | d) | Lucknow |

63. Recent classification of mango is given by:

| a) | Mukherjee and Majumdar | b) | Litz et al |
|---|---|---|---|
| c) | Kosterman and Bompard | d) | Swingle and Tanaka |

64. Which of the following Mangifera species is resistant to anthracnose?

| a) | Mangifera lagenifera | b) | Mangifera odorata |
|---|---|---|---|
| c) | Mangifera laurina | d) | Mangifera pajang |

65. Number of bee hives required for one hectare of litchi:
    a) 5-10 b) 10-15
    c) 15-20 d) 20-25
66. Which grape variety is commercially used for champagne making across the world?
    a) White Riesling b) Thompson Seedless
    c) Pinot Noir d) Pearl of Casaba
67. Flavouring compound in jamun is:
    a) Methyl salicylate b) Methyl anthranilate
    c) Isopentanol d) Methyl butyrate
68. Phyllanthoid branching habit is a characteristic feature of:
    a) Aonla b) Avocado
    c) Phalsa d) Karonda
69. Dichotomous branching habit is found in:
    a) Custard apple b) Loquat
    c) Phalsa d) Karonda
70. Lahsua belongs to .............family:
    a) Apocynaceae b) Tiliaceae
    c) Cappridaceaae d) Boraginaceae
71. Which of the following is a natural dwarf mutant of arecanut?
    a) VTLH-1 b) VTLH-2
    c) VTL-11 d) Hirehalli Dwarf
72. Which of the following is/are the best quality/ies cocoa?
    a) Criollo b) Forestero
    c) Trintario d) All the above
73. Oil palm mesocarp oil is rich in:
    a) Lauric acid b) Oleic acid
    c) Palmitic acid d) Linolenic acid
74. Pusa Lalima is a cross of:
    a) Amrapali x Sensation b) Amrapali x Lal Sundari
    c) Dashehari x Sensation d) Dashehari x Lal Sundari

75. Inverted bottleneck occurs in her .... when budded onto:

a) Ziziphus rotundifolia  b) Ziziphus paludosa
c) Ziziphus onepilea  d) Ziziphus nummularia

76. Which of the following is a progenitor of apple?

a) Mains oriental is  b) Mains b ace at a
c) Mains sylvestris  d) Malus sieversii

77. Which of the following is an apomictic species of apple?

a) Malus orientalis  b) Malus toringoides
c) Malus floribunda  d) Malus sieversii

78. Grade size of pitto apple is:

a) 45 mm  b) 50 mm
c) 55 mm  d) 60 mm

79. The method of heating soil by covering it with transparent polythene sheeting during hot periods to control soil borne diseases is known as:

a) Soil pasteurization  b) Soil treatment
c) Soil fumigation  d) Soil solarization

80. Grape variety having red colour both in skin and flesh is known as:

a) Teinturier grape  b) Coloured grape
c) Canning grape  d) None of these

81. Synthetic seed in banana was developed by:

a) NRC for Banana  b) BARC
c) IIHR  d) Jain Irrigation

82. For harvesting Delicious apple at right maturity, the Starch Pattern Index (SPI) should be:

a) 2.5/10  b) 3.5/10
c) 4.5/10  d) 6.5/10

83. 'Nishkant', a thorn less rootstock of rose was developed at:

a) IARI  b) IIHR
c) NBRI  d) IHBT

84. Seed rate of lawn is:

a) $2.5 g/m^2$  b) $25 g/m^2$
c) $250 g/m^2$  d) $2.5 kg/m^2$

85. Which of the following is not a leafless cactus?
    a) Aprocactus b) Cephalocereus
    c) Pereskia d) Mammilaria
86. Spores are used to propagate:
    a) Mosses b) Ferns
    c) Asparagus d) Bromeliads
87. Which of the following is used as an indicator plant for fluorine pollution?
    a) Ferns b) Tulip
    c) Gladiolus d) Iris
88. Basic principle of plant breeding is:
    a) Variation and selection b) Hybridization and selection
    c) Mutation and selection d) All of these
89. Optimum pH of holding solution should be:
    a) 2-3 b) 4-5
    c) 6-7 d) 7-8
90. Non-tunicate bulbs are common in:
    a) Tulip b) Narcissus
    c) Nerine d) Lilium
91. The largest producer of cut flowers in the world is:
    a) Israel b) USA
    c) Germany d) The Netherlands
92. Negative geotropism, a disorder of gladiolus occurs during:
    a) Harvesting b) Packing
    c) Transporting d) Selling
93. Which of the following is known as 'Nature in Miniature'?
    a) Italian garden b) Mughal garden
    c) Japanese garden d) Spanish garden
94. Calyx splitting of carnation is due to:
    a) Genetic factors b) Nutritional factors
    c) Environmental factors d) All of the above
95. Dieffenbachia is propagated by:
    a) Cane cutting b) Leaf cutting
    c) Air layering d) Root cutting

96. Fairy ring disease of lawn is caused by:
    a) Nematode b) Fungus
    c) Insects d) Parasites
97. Which annual can be grown more successfully in shade?
    a) Pansy b) Stock
    c) Cineraria d) Daisy
98. The scientific name of 'Sita ashok' is:
    a) Saraca indica b) Shorea robusta
    c) Bombax ceiba d) Mesua ferrea
99. Trimming of hedges is done when it attains the height of:
    a) 10 cm b) 15 cm
    c) 20 cm d) 25 cm
100. Who first started tissue culture in orchids?
    a) Morel b) Miller and Skoog
    c) Kogl d) Haberlandt
101. Aerial bulblets formed in the axil of leaves are known as:
    a) Aerial bulb b) Bulbils
    c) Cormels d) None of these
102. Orchid seeds are devoid of:
    a) Seed coat b) Cotyledon
    c) Endosperm d) Embryo
103. The 'Grandiflora roses' have been developed from a cross between:
    a) Hybrid Tea x Floribunda b) Hybrid T x Polyantha
    c) Hybrid polyantha x Tea rose d) Hybrid perpetual x Tea rose
104. First variety of 'Hybrid-T' rose is:
    a) La France b) First Prize
    c) La Paquerette d) Rodhte
105. Genetic male sterility in marigold is controlled by:
    a) Single dominant gene b) Single recessive gene
    c) Polygene d) None of these
106. A plant or plant parts composed of genetically different layers is known as:
    a) Chimera b) Mutant
    c) Hybrid d) Variant

107. How many pairs of contrasting characters of garden pea were selected by Mendel for his experiment?

a) 4 b) 7

c) 10 d) 14

108. Cytoplasmic male sterility is a common feature in:

a) Antirrhinum b) Primula

c) Petunia d) Ageratum

109. Which of the following is not a self-pollinated flowering annual?

a) Lupine b) Sweet pea

c) Saponaria d) Larkspur

110. In Japanese language, sand garden is known as:

a) Ryoanji b) Roji-niwa

c) Hira-niwa d) Soto-roji

111. Which of the following loose flowers covers highest area in India?

a) Marigold b) Jasmine

c) Chrysanthemum d) Tuberose

112. Isolation distance for certified seed production in carrot is:

a) 500 m b) 800 m

c) 1000 m d) 1600 m

113. 'Pillow' ,a physiological disorder of is caused due to calcium deficiency.

a) Cucumber b) Radish

c) Carrot d) Turnip

114. Starch content (%) of sweet potato tuber is:

a) 6 b) 16

c) 26 d) 0.6

115. Black leg in potato is caused by:

a) Fungus c) Bacteria

b) Virus d) MLOs

116. Solanum maglla, a wild species of potato is resistant to:

a) Frost b) Heat

c) Draught d) Flood

117. Is the progenitor form of the cultivated cucumber.

a) *Cucumis sativus* var. hardwickii b) Cucumis melo

c) Cucumis angurina d) Cucumis lantaus

118. Non-lobed leaf character is used as single recessive marker gene for hybrid seed production in:
   a) Watermelon  b) Muskmelon
   c) Cucumber  d) None of these
119. Swarna Rekha is an improved variety of:
   a) Pointed gourd  b) Snake gourd
   c) Ivy gourd  d) Bottle gourd
120. Seed rate of bitter gourd is:
   a) 5kg/ha  b) 15kg/ha
   c) 25kg/ha  d) 35kg/ha
121. Butter bean is the nick name of:
   a) Dolichus bean  b Faba bean
   c) Lima bean  d) French bean
122. Hypocotyl necrosis is a physiological disorder of:
   a) French bean  b) Cluster bean
   c) Cowpea  d) Runner bean
123. Ivy gourd is propagated through:
   a) Tuberous root  b) Seed
   c) Sprouted fruit  d) Stem cutting
124. Great Lakes variety of lettuce belongs to .........group:
   a) Crips head  b) Butter head
   c) Romaine  d) Leafy or bunching
125. The ploidy level of sweet potato is:
   a) Diploid  b) Tetraploid
   c) Hexaploid  d) Octaploid
126. Heat unit is used to determine the maturity of:
   a) Garden pea  b) Cowpea
   c) French bean  d) Cluster bean
127. Which of the following cucurbits do not have trailing habit?
   a) Winter squash  b) Summer squash
   c) Snap melon  d) Pumpkin
128. Which of the following vegetables is used as 'trap crop'?
   a) Tomato  b) Brinjal
   c) Okra  d) Cucumber

129. Melons for distant marketing are harvested at:

a) Green mature stage b) Half-slip stage

c) Full-slip stage d) Immature green stage

130. The gene responsible for nematode resistance in tomato is:

a) Mi b) Ni

c) Ri d) Ki

131. Which of the following is a dual purpose variety of fenugreek?

a) Co-1 b) Hisar Sonali

c) Pant Ragini d) Lam selection-1

132. Which of the following varieties of ginger contains highest oleoresin'

a) Rio-de- Janeiro b) Wynad Manantody

c) Suprabha d) Suruchi

133. The concentration of boron (B) used to increase fruit set in bottle gourd is:

a) 0.3 ppm b) 3 ppm

c) 30 ppm d) 300 ppm

134. National Facility for Virus Diagnosis and Quality Control of Tissue Culture raised plants is located at:

a) IARI b) IIHR

c) NBRI d) IHBT

135. The plant growth regulator used to overcome dormancy and enhance germination is:

a) Kinetta b) $GA_3$

c) Zearin d) ABA

136. Pseado-bulbs are used to propagate:

a) Lilhan b) Tuberose

c) Orchids d) Tulip

137. The biggest international flower market is located at:

a) Washington b) Singapore

c) London d) Aalsmeer

138. Which is a very serious pest of chrysanthemum?

a) Aphid b) Red spider mite

c) Heliothis d) Thrips

139. FIorissant-500 is a/an:

a) Fungicide b) Flower preservative

c) Software d) Insecticide

140. $CO_2$ concentration in greenhouse for rose growing should be:

a) 500 ppm b) 4000-6000 ppm

c) 1000-3000 ppm d) 500-1000 ppm

141. The major disadvantage of transportation of vacuum cooled cut flower is:

a) Increase in respiration rate b) Loss of moisture

c) Yellowing of leaves d) Increase in ethylene production

142. Offsets are very common for propagation of :

a) Agave b) Aloe

c) Pandanus d) All of these

143. Marigold has originated from?

a) USA b) Mexico

c) West Indies d) South Africa

144. Pusa Tara is cultivar of:

a) Marigold b) Cosmos

c) Coreopsis d) Aster

145. Major problematic weed in lawn is:

a) Cyperus rotundas b) Convolvulus arvensis

c) Phyllantus nirvri d) Parthenium histerophorits

146. For growing cacti, which type of pots are most suitable?

a) Cement b) Plastic

c) Concreate d) Brass

147. Hypobaric storage is also knows as:

a) Low temperature storage b) Low pressure storage

c) Cold storage d) CA storage

148. Which of the following is the hardiest annual?

a) Digitalis b) Stock

c) Antirrhinum d) Aster

149. Netted varieties of muskmelon are known as:

a) Honey Dews b) Casabas

c) Cantaloupes d) None of these

150. Multigerm seed is found in case of:
   a) Onion b) Carrot
   c) Tomato d) Beetroot
151. Limp neck is a physiological disorder of:
   a) Rose b) Gladiolus
   c) Tuberose d) Tulip
152. Chilling injury is common in:
   a) Cattleya b) Cymbidiuna
   c) Vanda d) PaphiopedikHn
153. Which of the following isless sensitive to ethylene:
   a) Chrysanthemum b) Kalanchoe
   c) Begonia d) Beloperone
154. Which garden is regarded as 'Genesis of gardeaiag'
   a) Hampshire b) Osaka
   c) Eden d) Briodavan
155. International Bougainvillea Registration Centre is situated at:
   a) LARI b) IIHR
   c) NBRI d) TNAU
156. Betalians is an active compound found in:
   a) Rose b) Ctous
   c) Avocado d) Beetroot
157. Diploid apogamy is found in:
   a) Allium spp. b) Knol-khol
   c) Okra d) Cluster bean
158. Which of the following French bean varieties is developed through mutation?
   a) Pusa Parvati b) Phule Surekha
   c) Pant Affinpaxn d) Alia Komal
159. Sporophytic self-incompatibility is observed in:
   a) Onion b) Sweet potato
   c) Tomato d) Cole crops
160. Black rot resistant variety of cauliflower is:
   a) Pusa Deepali b) Pusa Katki
   c) Pusa Himjyoti d) Pusa Shubhra

161. Anthesis occurs during evening hours in:
    a) Bottle gourd    b) Muskmelon
    c) Cucumber    d) None of these

162. Which of the following is a C4 plant:
    a) Onion    b) Amaranth us
    c) Tomato    d) Beetroot

163. Rutin' and vitamin B are associated with which vegetable crop?
    a) Tomato    b) Brinjal
    c) Potato    d) Chilli

164. Rhubarb is a native of:
    a) Europe    b) Mediterranean region
    c) Siberia    d) Brazil

165. The mineral element that plays an important role in translocation of sugar is:
    a) Ca    b) Mg
    c) Cu    d) K

166. Iron (Fe) deficiency in vegetables causes:
    a) Reduction of internode length
    b) Interveinal chlorosis of younger leaves
    c) Cupping of leaves
    d) Anthocyanin pigmentation in leaves

167. Solasodine, a bitter glycoalkaloid is found in:
    a) Brinjal    b) Potato
    c) Tomato    d) Chilli

168. Cheratin is isolated from:
    a) Bottle gourd    b) Bitter gourd
    c) Chilli    d) Cauliflower

169. In the development of hybrids, availability of male sterile lines plays significant role in:
    a) Sweet potato    b) Onion
    c) Brinjal    d) Potato

170. Akashin is a physiological disorder of:
    a) Radish    b) Carrot
    c) Beetroot    d) Turnip

171. Which of the following exhibits very low respiration rate?
    a) Onion b) Beans
    c) Tomato d) Spinach
172. Cluster bean belongs to .........genera:
    a) Cyamopsis b) Phascolns
    c) Crotolaria d) Vigna
173. Cryoprcservation is done in liquid nitrogen at:
    a) -160°C b) -196°C
    c) -96°C d) -126°C
174. Which of the following plant parts is considered free from virus?
    a) Apical meristem b) Petiole
    c) Midrib d) Internode
175. Kufri Jyoti variety of potato is resistant to:
    a) Cyst nematode b) Early blight
    c) Black leg d) Late blight
176. Arka Abhay is a variety of:
    a) Brinjal b) Okra
    c) Tomato d) Watermelon
177. The part of celery used as salad:
    a) Fruit b) Flower
    c) Shoot d) Petiole
178. For low sugar content, potato tubers are stored at:
    a) 5°C b) 10°C
    c) 15°C d) 20°C
179. Vacuum cooling is most suitable for:
    a) Fruit vegetables b) Tuber vegetables
    c) Leafy vegetables d) Root vegetables
180. Tomato ketchup is preserved by:
    a) KMS b) Sodium benzoate
    c) Citric acid d) Sugar
181. Solidity is the maturity index for:
    a) Root vegetables b) Seed vegetables
    c) Leafy vegetables d) Cucurbits

182. In cucumber, chilling injury symptoms occurr at:
   - a) <7°C
   - b) 10°C
   - c) 13°C
   - d) 15°C

183. Which of the following does not produce enzymes?
   - a) Fungus
   - b) Bacteria
   - c) Virus
   - d) MLOs

184. Black Wattle (medicinal plant) is harvested after years of growth.
   - a) 5
   - b) 10
   - c) 15
   - d) 20

185. The effects of 'Genetic drift' are much easier to see in:
   - a) Small population
   - b) Large population
   - c) Random mating population
   - d) None of these

186. Crossing over takes place in:
   - a) Laptotene
   - b) Pachytene
   - c) Diplotcne
   - d) Zygotene

187. Which of the following is a rich source of iodine content?
   - a) Turnip
   - b) Cabbage
   - c) Onion
   - d) Okra

188. In okra, which character shows pleotropic effect?
   - a) Resistant to YVMV
   - b) Resistant to powdery mildew
   - c) Calyx and petal vein color
   - d) Leaf lobation

189. Diverse cytoplasm can be fixed by:
   - a) Protoplast fusion
   - b) Somaclonal variation
   - c) Hybridization
   - d) Mutation

190. Which of the following plays an important role in opening and closing of stomata?
   - a) N
   - b) P
   - c) K
   - d) Ca

191. Which of the following species of arecanut is closely related to cultivated type?
   - a) Areca triandra
   - b) Areca multifida
   - c) Areca concinna
   - d) Areca whitfordii

192. The major breeding objective of oil palm is to improve:
    a) Oil quality b) Yield
    c) Tolerance to pest and disease d) Reduce height
193. The bitter taste of black pepper is because of:
    a) Terpenine b) Piperine
    c) Pipermenthene d) Pentanol
194. Konkan Sugandha is a variety of:
    a) Nutmeg b) Allspice
    c) Kokum d) Clove
195. Which of the following varieties of cardamom is resistant to rhizome rot?
    a) Vijeta b) ICRI-4
    c) Mudigere-1 d) Suvashini
196. The turmeric variety that contains highest oleoresin is:
    a) Roma b) Suroma
    c) Prabha d) Prathibha
197. Mass selection is a method of:
    a) Plant breeding b) Population improvement
    c) Mutation d) Progeny testing
198. An alkaloid present in Aloe vera is:
    a) Aloin b) Codein
    c) Solasodine d) Carvone
199. The active ingredient of lavender oil is:
    a) Linalool b) Methyl chavicol
    c) Farnesol d) Geraniol
200. Seed rate (rhizome) required for one hector of ginger planting is:
    a) 1000-1400 kg b) 1500-1800 kg
    c) 1800-2200 kg d) 2500 kg

## Answers Key

| | | | | | | | | | | | | | |
|---|---|---|---|---|---|---|---|---|---|---|---|---|---|
| 1. | b) | 2. | d) | 3. | b) | 4. | b) | 5. | a) | 6. | c) | 7. | a) |
| 8. | d) | 9. | b) | 10. | d) | 11. | c) | 12. | d) | 13. | c) | 14. | c) |
| 15. | d) | 16. | b) | 17. | a) | 18. | d) | 19. | c) | 20. | a) | 21. | d) |
| 22. | c) | 23. | a) | 24. | d) | 25. | a) | 26. | a) | 27. | a) | 28. | d) |

| | | | | | | |
|---|---|---|---|---|---|---|
| 29. b) | 30. b) | 31. d) | 32. a) | 33. d) | 34. b) | 35. b) |
| 36. a) | 37. c) | 38. b) | 39. b) | 40. c) | 41. c) | 42. b) |
| 43. b) | 44. b) | 45. a) | 46. c) | 47. c) | 48. b) | 49. c) |
| 50. b) | 51. b) | 52. b) | 53. d) | 54. c) | 55. b) | 56. c) |
| 57. c) | 58. c) | 59. c) | 60. b) | 61. b) | 62. c) | 63. c) |
| 64. c) | 65. c) | 66. a) | 67. b) | 68. a) | 69. d) | 70. d) |
| 71. d) | 72. a) | 73. c) | 74. c) | 75. d) | 76. d) | 77. b) |
| 78. c) | 79. d) | 80. a) | 81. b) | 82. c) | 83. b) | 84. a) |
| 85. c) | 86. b) | 87. c) | 88. a) | 89. b) | 90. d) | 91. d) |
| 92. c) | 93. c) | 94. d) | 95. a) | 96. b) | 97. c) | 98. a) |
| 99. b) | 100. a) | 101. b) | 102. c) | 103. a) | 104. a) | 105. b) |
| 106. a) | 107. b) | 108. c) | 109. d) | 110. a) | 111. b) | 112. b) |
| 113. a) | 114. b) | 115. c) | 116. b) | 117. a) | 118. a) | 119. a) |
| 120. a) | 121. c) | 122. a) | 123. d) | 124. a) | 125. c) | 126. a) |
| 127. b) | 128. c) | 129. b) | 130. a) | 131. a) | 132. a) | 133. b) |
| 134. a) | 135. b) | 136. c) | 137. d) | 138. a) | 139. b) | 140. c) |
| 141. b) | 142. d) | 143. b) | 144. c) | 145. a) | 146. b) | 147. b) |
| 148. a) | 149. c) | 150. d) | 151. a) | 152. d) | 153. a) | 154. c) |
| 155. a) | 156. d) | 157. a) | 158. a) | 159. d) | 160. d) | 161. a) |
| 162. b) | 163. d) | 164. c) | 165. d) | 166. b) | 167. a) | 168. b) |
| 169. b) | 170. a) | 171. a) | 172. a) | 173. b) | 174. a) | 175. d) |
| 176. b) | 177. d) | 178. b) | 179. c) | 180. b) | 181. c) | 182. a) |
| 183. c) | 184. b) | 185. a) | 186. b) | 187. d) | 188. c) | 189. a) |
| 190. c) | 191. a) | 192. a) | 193. b) | 194. a) | 195. b) | 196. d) |
| 197. b) | 198. a) | 199. a) | 200. b) | | | |

# 11

# ICAR – SRF Horticulture Exam – 2016

1. Alphabet Inc. (marketed as Alphabet) is the parent company of:

   a) Google  b) Yahoo

   c) Microsoft  d) Apple

2. Human Rights Day is celebrated annually across the world on ................ every year.

   a) 10 October  b) 10 November

   c) 10 December  d) 10 January

3. In India, National Consumer Day is celebrated on:

   a) 15 March  b) 24 December

   c) 23 September  d) 21 April

4. The angle between the minute hand and the hour hand of a clock when the time is 8.30 is:

   a) 60°  b) 75°

   c) 80°  d) 105°

5. Fish eggs are also referred to as:

   a) Roe  b) Tofu

   c) Balut  d) Pub

6. Which of the following is/are function/s of NABARD?

   a) It acts as an apex body for meeting the credit needs of all types of agricultural and rural development

   b) It provides short-term, medium-term and long-term credit to SCBs, LDBs, RRBs and approved financial institutions.

   c) It has the responsibility of inspecting co-operative banks and RRBs.

   d) All of these

7. The 'Statue of Unity' is located at:

a) West Bengal b) Gujrat
c) Maharashtra d) Uttar Pradesh

8. National Innovations on Climate Resilient Agriculture (NICRA) was launched in:

a) 2011 b) 2012
c) 2013 d) 2014

9. Which of the following is a hybrid variety of arecanut?

a) VTL-17 b) VTL-75
c) VTL-11 d) VTLH-1

10. Which of the following varieties of pineapple is suitable for canning?

a) Queen b) Kew
c) Mauntius d) Jaldhoop

11. Oil palm comes into bearing in ...........year after planting.

a) 5 b) 10
c) 15 d) 20

12. Zero energy cool chamber was developed by:

a) M. K. Rai and R. N. Singh b) S. K. Roy and D. S. Khurdiya
c) R. P. Roy and D. K. Khurana d) None of these

13. MM-111 rootstock of apple is classified as:

a) Dwarf b) Semi dwarf
c) Semi vigorous d) Vigorous

14. Major constraint in date palm cultivation under north-Indian condition is:

a) High temperature during ripening
b) Winter frost
c) Rains during ripening
d) Fungal diseases

15. Cashew Nut Shell Liquid (CNSL), a naturally occurring phenol is obtained from:

a) Epicarp b) Mesocarp
c) Endocarp d) Endosperm

16. Kalipak is prepared from:

a) Tender nut b) Ripe nut
c) Dried nut d) None of these

17. Most popular hybrid clones of rubber in India is:
    a) RRIM-600 b) RRIM-628
    c) RRII-105 d) RRIM-703
18. Controlled Atmosphere Storage was invented by:
    a) Kidd and West b) Gane
    c) James Harrison d) Wade
19. What should be the height of support system in head system of training in grape?
    a) 1.5 m b) 1.8 m
    c) 2.5 m d) 3 m
20. How many plants can be accommodated in spindle bush system of apple?
    a) 2000 plants/ ha. b) 3000 plants/ ha.
    c) 5000 plants/ ha. d) 10000 plants/ha.
21. Which of the following fruit crops is known as 'Smiling nut'?
    a) Walnut b) Cashew nut
    c) Pistachio nut d) Pecan nut
22. Which of the following varietites of grape is suitable for sparkling wine making?
    a) Thompson Seedless b) Beauty Seedless
    c) Pinot Noir d) Perlette
23. TSS/acid ratio of Kinnow mandarin at the time of harvesting should be:
    a) 8:1-10:1 b) 10:1-12:1
    c) 12:1-14:1 d) 14:1-16:1
24. The concept of 'Photoperiodism' was given by:
    a) Lysenko b) Garner and Allard
    c) Kidd and West d) F. W. Went
25. Which of the following is known as 'self- peeling banana'?
    a) Musa paterita b) Musa ingens
    c) Musa basjoo d) Musa velutina
26. Which of the following is a hybrid variety of Ziziphus mauritiana :
    a) That Bhubhraj b) Thar Sevika
    c) Goma Kirti d) ZG-2

27. The AEZ (Agri Export Zone) identified for Kesar mango is:
    a) Andhra Pradesh and Maharashtra
    b) Gujrat and Maharashtra
    c) Maharashtra and Karnataka
    d) Gujrat and Uttar Pradesh
28. TSS of Thompson Seedless grape at the time of maturity is:
    a) 12-14°B b) 14-16°B
    c) 16-18°B d) 18-22°B
29. Which of the following is a hybrid variety of arecanut?
    a) VTL-17 b) VTL-75
    c) VTL-11 d) VTLH-1
30. Which of the following varieties of pineapple is suitable for canning?
    a) Queen b) Kew
    c) Mauritius d) Jaldhoop
31. Oil palm comes into bearing........................year after planting.
    a) 5 b) 10
    c) 15 d) 20
32. Zero energy cool chamber was developed by:
    a) M.K. Rai and R.N. Singh b) S.K. Roy and D.S. Khurdiya
    c) R.P. Roy and D.K. Khurana d) None of these
33. MM-111 rootstock of apple is classified as:
    a) Dwarf b) Semi dwarf
    c) Semi vigorous d) Vigorous
34. Major constraint in date palm cultivation under north-Indian condition is:
    a) High temperature during ripening
    b) Winter frost
    c) Rains during ripening
    d) Fungal diseases
35. Cashew Nut Shell Liquid (CNSL) naturally occurring phenol is obtained from:
    a) Epicarp b) Mesocarp
    c) Endocarp d) Endosperm

36. Kalipak is prepared from:

a) Tender nut b) Ripe nut

c) Dried nut d) None of these

37. Most popular hybrid clone of rubber in India is:

a) RRIM-600 b) RRIM-628

c) RRII-105 d) RRIM-703

38. Controlled Atmosphere Storage was invented by:

a) Kidd and West b) Gane

c) James Harrison d) Wade

39. What should be the height of support system in head system raining in grape?

a) 1.5 m b) 1.8 m

c) 2.5 m d) 3 m

40. How many plants can be accommodated in spindle bush system of apple?

a) 2000 plants/ ha. b) 3000 plants/ ha.

c) 5000 plants/ ha. d) 10000 plants/ha.

41. Partial pinching of crown to improve fruit size and shape in pineapple is done at:

a) 25 days after fruit set b) 35 days after fruit set

c) 45 days after fruit set d) 55 days after fruit set

42. Which of the following is a Peach x AImond hybrid?

a) Lovell b) GF-557

c) Mariana d) Nemagaurd

43. Parental percentage of 'Plutos' is:

a) 50% plum and 50% apricot b) 25% plum and 75% apricot

c) 75% plum and 25% apricot d) None of these

44. Total sugar content of fresh fig fruit is:

a) 5% b) 10%

c) 15% d) 20%

45. Which of the following techniques is used for detection of viruses in citrus?

a) RT-PCR b) ELISA

c) ISSR d) PCR

46. Sex form of nutmeg is:
    a) Monoecious b) Dioecious
    c) Hermaphrodite d) None of these
47. Which of the following is a very shallow rooted vegetable?
    a) Spinach b) Beet root
    c) Cabbage d) Onion
48. Which of the following countries has the highest productivity of vegetables in the world?
    a) Spain b) Indonesia
    c) USA d) Thailand
49. The recovery (%) of black pepper is about percent of ripe berries.
    a) 25% b) 28%
    c) 33% d) 38%
50. Majority of coconut roots are concentrated within a diameter of:
    a) 50 cm b) 2 m
    c) 4m d) 6 m
51. In which of the following fruit crops, cell division continues till maturity in fruit development:
    a) Citrus b) Strawberry
    c) Persimmon d) Avocado
52. Anthocyanin pigment in apple fruits is found in the form of:
    a) Lycopene b) Cynaidin-3-glucoside
    c) p-carotene (Xanthophyll) d) 5-dehydroshikimic acid
53. Which of the following races of avocado has the highest oil content?
    a) Mexican b) Guatemalan
    c) West Indian d) Phalsa
54. Double sigmoid growth curve is observed in:
    a) Apple b) Pear
    c) Peach d) Almond
55. Internal break down in pomegranate occurs in:
    a) Ambe bahar b) Mrig bahar
    c) Hasth bahar d) All of these

56. Which of the following is not a Ca related physiological disorder?
    a) Fruit cracking of pomegranate b) Bitter pit of apple
    c) Calyx end rot of persimmon d) Internal fruit necrosis of aonla
57. During fruit ripening, usually organic acids decline with the exception in:
    a) Mango b) Tomato
    c) Guava d) Banana
58. Degreening in citrus fruits is done with the application of:
    a) GA3 b) Auxins
    c) Ethephon d) Cytokinins
59. Which of the following microorganisms causes fatal poisoning in canned fruits and vegetables?
    a) Aspergillus flavus b) Penicillium digitatum
    c) Clostridium botulinum d) Rhizoctonia solani
60. Which of the following is the second stage of syconium development in fig?
    a) Tissue softening stage b) Sugar accumulation stage
    c) Quiescence stage d) Starch accumulation stage
61. NDVI stands for:
    a) Normalized Difference Vegetation Index
    b) Nationalized Difference Vegetation Index
    c) Normalized Difficult Vegetation Index
    d) Nationalized Difficult Vegetation Index
62. Which of the following edible coating is commercially recommended for strawberry?
    a) Bilayer b) Zein
    c) Chitosan d) Whey
63. The thickness of edible coating applied to fruit should be less than:
    a) 0.3 mm b) 0.6 mm
    c) 0.9 mm d) 1.2 mm
64. Size of tapping panel in budded plants of rubber is:
    a) 50 cm girth and 50 cm height b) 100 cm girth and 100 cm height
    c) 50 cm girth and 125 cm height d) 100 cm girth and 125 cm height

65. The speed of grafting fruit trees strongly depends on practice, generally known as:
    a) Blue finger b) Red finger
    c) Green finger d) Nursed finger
66. Gene silencing in fruit crops is used for:
    a) DNA inhibition b) Protein inhibition
    c) Enzyme inhibition d) None of these
67. In tea, the table is lowered down to convenient height of 55-70 cm to eliminate the thin and weaker laterals. The process is known as:
    a) Light pruning b) Medium pruning
    c) Hard pruning d) Rejuvenation pruning
68. Forquette emerge from the thupon in cocoa from a height of:
    a) 1-2 m b) 2-3 m
    c) 3-4 m d) 4-5 m
69. Most widely grown species of coffee is:
    a) *Coffea arabica* b) *Coffea canephora*
    c) *Coffea liberica* d) All of these
70. Coffee leaves need only .................. of the available sunlight to function at top performance.
    a) 20-25% b) 40-45%
    c) 60-65% d) 80-85%
71. Which instrument is used to measure the amount of water in air?
    a) Psychrometer b) Pyranometer
    c) Pyrheliometer d) Porometer
72. The variations seen in plants that have been produced by plant tissue culture are known as:
    a) Chimeric variation b) Somaclonal variation
    c) Mutated variation d) None of these
73. Tetrapacks are used for:
    a) Aseptic packaging of food products
    b) Aseptic canning of food products
    c) Thermal processing of food products
    d) Aseptic processing of food products

74. Grape guard is made of:
    a) KMS + Citric acid  b) KMnO4+ Citric acid
    c) KMS + sodium benzoate  d) KMnO4 + sodium benzoate
75. Coriander is harvested at which stage?
    a) Yellowing of leaves  b) Drying of seeds
    c) 50% seeds gets brown  d) None of these
76. Cardamom has originated from:
    a) Tropical Africa  b) North America
    c) China  d) India
77. Which of the following countries is the largest producer of cardamom?
    a) Guatemala  b) Vietnam
    c) Sri Lanka  d) India
78. Clove is propagated through:
    a) Rhizome  b) Corm
    c) Mother clove  d) Cutting
79. Which of the following agencies regulates the certification of organic products?
    a) Spice board  b) APEDA
    c) ICAR  d) Indocert
80. The economic part of thyme is:
    a) Leaves  b) Roots
    c) Flowers  d) Whole plant
81. Defective broken bits of coffee are known as:
    a) Triage  b) Wastage
    c) Blacks  d) Splits
82. When performed, the root initials cause swelling at nodal region referred to as:
    a) Nodules  b) Blisters
    c) Burr knots  d) Galls
83. Under multi-storied intercropping with coconut, the tree crop interaction is:
    a) Complementary  b) Competitive
    c) Allelopathic  d) Supplementary

84. Shot berry is a serious problem in ........grape variety:
   a) Perlette b) Delight
   c) Gold d) Cardinal
85. The banana variety used for chip making is:
   a) Monthan b) Cavendish
   c) Nendran d) Poovan
86. The ideal temperature for hydro cooling of mango fruits is:
   a) 3-6% b) 7-10%
   c) 12-15% d) 18-20%
87. The easiest and most commonly used method of breeding in unexploited fruit crops is:
   a) Clonal selection b) Mutation breeding
   c) Hybridization d) Polyploidy breeding
88. Since fruits crops are highly heterozygous, the selection after hybridization has to initiate in:
   a) F1 generation b) F2 generation
   c) F3 generation d) F4 generation
89. Abortion of embryo before maturation resulting in seedlessness is a serious problem in improvement of which fruit crop?
   a) Litchi b) Mango
   c) Grape d) Banana
90. Compact Lambert variety of cherry is:
   a) Self-fruitful b) Self-unfruitful
   c) Self-fertile d) Self-sterile
91. Mutation is a sudden heritable change in the genetic makeup of an organism. It isa:
   a) Multicellular event b) Organ event
   c) Single cell event d) All of the above
92. Cam plants which die after once flowering and fruiting are known as:
   a) Monocarpic b) Bicarpic
   c) Polycarpic d) Monoecious
93. Nematode resistant rootstock of fig is known as:
   a) *Ficus infectoria* b) *Ficus religiosa*
   c) *Ficus glomerata* d) *Ficus benghalensis*

94. Pear decline is related to which of the following?
   a) Pear psylla  b) Nutrients deficieny
   c) Pathogen attack  d) All of the above

95. Which type of banana bears only female flower buds?
   a) French plantain  b) Horn plantain
   c) False horn  d) French horn

96. Lemon turns bitter on:
   a) Reduction  b) Oxidation
   c) Neutralization  d) None of these

97. Which of the following is commonly known as the Japanese mint?
   a) *Mentha arvensis*  b) *Mentha piperita*
   c) *Mentha spicata*  d) *Mentha cardiaca*

98. The seed rate of lemon grass is:
   a) 2-4 kg/ha  b) 4-6 kg/ha
   c) 6-8 kg/ha  d) 8-10 kg/ha

99. Vanilla is harvested at which stage?
   a) Fully ripe stage  b) Immature stage
   c) Tender stage  d) Ripe stage

100. 'Indian Ginseng' is the common name of:
   a) *Rauwolfia serpentina*  b) *Hemidesmus indicus*
   c) *Azadiracta indica*  d) *Withania somnifera*

101. Sir Edwin Lutyens was a/an:
   a) Architect  b) Breeder
   c) Singer  d) Horticulturist

102. Who was famous a French garden designer?
   a) L. Brown  b) H. Hoare
   c) Le Notre  d) C. Lorrain

103. Roshanara Park at New Delhi is a style garden.
   a) Mughal  b) Japanese
   c) Italian  d) English

104. Rajat Rekha and Swarna Rekha are the mutant cultivar of:
   a) Dahlia  b) Tuberose
   c) Rose  d) Portulaca

105. Inheritance of pigments is controlled by .............................. gene action.

a) Additive b) Dominance
c) Epistasis d) None of these

106. Change from mutant allele to wild type is known as:

a) Forward mutation b) Reverse mutation
c) Somatic mutation d) Nuclear mutation

107. Normal relative humidity in greenhouse should be:

a) 20-40% b) 40-60%
c) 60-80% d) 80-100%

108. Which species of *Tagetes* is commercially grown for oil extraction?

a) *T. tenuifolia* b) *T. erecta*
c) *T. patula* d) *T. minuta*

109. Which species of jasmine has the highest recovery of oil?

a) *Jasminuni aitriciilatum* b) *Jasminuni grandiflorum*
c) *Jasminum sambac* d) *Jasminum multiflorum*

110. Japanese arrangement of flowers is known as:

a) Ikebana b) Tatebana
c) Jiyubana d) Moribana

111. Which of the following is not a principle of landscaping?

a) Simplicity b) Balance
c) Line d) Rhythm

112. Seed rate of a lawn is:

a) $2.5g/m^2$ b) $25g/m^2$
c) $250g/m^2$ d) $2.5\ kg/m^2$

113. Which of the following is suitable to for saline soils?

a) *Casuarina equinetffolia* b) *Pinus roxburghii*
c) *Salix babylonica* d) *Grevillea robusta*

114. Which annual can be grown more successfully in shade?

a) Pansy b) Stock
c) Cineraria d) Daisy

115. Which of the following is a common oxygenator?

a) Elodea b) Crow foot
c) Victoria d) Typha

116. Queen Elizabeth and Lincoln are the common varieties of:

a) Rose b) Dahlia

c) Carnation d) Marigold

117. *Jacaranda acutifolia* produces flowers.

a) White b) Yellow

c) Red d) Blue

118. Flowering in *Cassia fistula* occurs during:

a) July-November b) November-February

c) February-April d) April-June

119. Male sterility is common in:

a) Petunia b) Marigold

c) Antirrhinum d) All of these

120. Lucky star is a fragrant cultivar of:

a) Rose b) Tuberose

c) Gladiolus d) Marigold

121. Which of the following is grown in rough twin scaling?

a) Gladiolus b) Tuberose

c) Tulip d) Lilium

122. The concept of 'Bio-aesthetic planning' was first propounded by:

a) Persy Lancaster b) Lancelot Hogbean

c) William Robinson d) M.S. Randhawa

123. Dr. H.B. Singh is a cultivar of:

a) Hibiscus b) China aster

c) Bougainvillea d) Gerbera

124. Mohini is a ..................... colour variety of rose.

a) Red b) Blue

c) Yellow d) Chocolate brown

125. The notable contribution of NBRI, Lucknow in floriculture is the development of number of cultivars in:

a) Chrysanthemum b) Jasmine

c) Marigold d) Dahlia

126 .................. is/are suitable for separation of dry flowers.

a) *Helichrysum* b) *Helipterum*

c) *Limonium* d) All of these

127. Bud drop or bud blast is a physiological disorder of:
   a) Orchids b) Gladiolus
   c) Alstromeria d) Lilium
128. *Ruscus* is a genus of:
   a) Temperate indoor plants b) Tropical shrubs
   c) Arid trees d) Alpines
129. The term 'Keikis' is related to:
   a) Ferns b) Lilium
   c) Orchids d) Chrysanthemum
130. Russelia juncea is suitable for:
   a) Sand garden b) Water garden
   c) Rock garden d) Sunken garden
131. 'Casa Grande' variety of chrysanthemum belongs to group:
   a) Incurved b) Reflexed
   c) Anemone d) Pompon
132. Classification of chrysanthemum based on temperature requirement was given by:
   a) Cathey b) Hogbean
   c) Went d) Hooker
133. BSA stands for:
   a) Bulked Segregant Assay b) Bulked Segregant Analysis
   c) Bio Segregant Analysis d) None of these
134. Bulbils are not formed in which of the following species of *Lilium*?
   a) *L. tigrinum* b) *L. bulbiferum*
   c) *L. longiflorum* d) *L. wallichianum*
135. Which of the following is the major carotenoid in the flowers of marigold?
   a) Lutein b) Lycopene
   c) Zeaxanthin d) Carotene
136. Bird of paradise (Strelitzia reginae) is native to:
   a) New Zealand b) South Africa
   c) Japan d) Germany
137. Which of the following is a foliage annual?
   a) Kochia b) Aster
   c) Cosmos d) Salvia

138. Carnation is classified as:
   a) Short day plant b) Long day plant
   c) Day neutral plant d) Long short day plant
139. Which of the following is a strongest senescence stimulator?
   a) BA b) ABA
   c) AOA d) PBA
140. Scoring is very common in which of the following bulbous plant?
   a) Lilium b) Gladiolus
   c) Tulip d) Begonia
141. Which of the following breeding methods is most successful in ornamental plants?
   a) Pure line selection b) Mass selection
   c) Mutation d) Hybridization
142. Quilling of florets is a common disorder of:
   a) Chrysanthemum b) Carnation
   c) Dahlia d) Tulip
143. Pseudo bulbs are commonly used to multiply:
   a) Orchids b) Gladiolus
   c) Tulip d) Tuberose
144. Mean, median and mode are equal in:
   a) Normal distribution b) Poisson distribution
   c) Binomial distribution d) None of these
145. Source of blue pigmentation in roses are found in .......... cultivar:
   a) Sonia b) Samba
   c) Blue Moon d) Bhim
146. Seed viability is determined by:
   a) Direct germination b) Excised embryo
   c) Tetrazolium test d) All of these
147. Which of the following is most suitable for hanging basket?
   a) Portulaca b) Cosmos
   c) Rudbeckia d) Aster
148. ELISA test is done to find out ...................... diseases.
   a) Fungal b) Bacterial
   c) Viral d) All of these

149. Diploid chromosome number of rose is:

a) 10 b) 14
c) 18 d) 20

150. Diocey is very common in:

a) Asparagus b) Tulip
c) Petunia d) Aster

151. Ty-1 gene is associated with:

a) Tomato b) Chilli
c) Cauliflower d) Cucumber

152. Which of the following crop has cn=24?

a) Tomato b) Ash gourd
c) Cucumber d) Carrot

153. 'Gold Fleck' which is caused due to excess of Ca oxalate and low K:Ca ratio occurs in:

a) Potato b) Carrot
c) Cabbage d) Tomato

154. Which of the following is/are indigenous vegetable(s)?

a) Pointed gourd b) Lablab bean
c) Brinjal d) All of these

155. Production of lycopene in tomato is adversely hampered at ................ temperature.

a) >20°C b) >25°C
c) >30°C d) >35°C

156. Which of the following species of tomato is resistant to insect/pests?

a) *S. chilense* b) *S. hirsutum*
c) *S. cheesmani* d) *S. parviflorum*

157. Which of the following *Solanum* species is tolerant to drought?

a) *Solanum peruvianum* b) *Solanum cheesmani*
c) *Solanum galapagense* d) *Solanum penelli*

158. Which of the following species of brinjal is used in grafting?

a) *Solanum sisymbriifolium* b) *Solanum torvum*
c) *Solanum gilo* d) *Solanum khasianum*

159. Red colour of onion is due to the presence of:

a) Lycopene b) Anthocyanin

c) Quercetin d) Xanthophyll

160. Which of the following is an ancestor species of watermelon?

a) *Citrullus vulgaris* b) *Citrullus colocynthi*

c) *Citrullus ecirrhosus* d) *Citrullus naudinianus*

161. Commonly used sprout suppressant in potato is:

a) EDB b) Thiourea

c) CIPC d) Biphenyl

162. Synthetic seed in vegetables is produced through:

a) Ovuje culture b) Callusvculture

c) Meristem culture d) Somatic embryogenesis

163. Potato is transformed with protein synthesizer gene from:

a) Fenugreek b) Spinach

c) Beet leaf d) Amaranthus

164. The chromosome number (2n=) of French bean is:

a) 18 b) 20

c) 22 d) 24

165. Optimum temperature required for vernalization in temperate vegetables is:

a) <4°C b) <7°C

c) <10°C d) <15°C

166. Andromonecius sex form is present in:

a) Watermelon b) Muskmelon

c) Longmelon d) Round melon

167. Acidic soil encourages:

a) Clubroot in crucifers b) Black leg in crucifers

c) Verticillium wilt in solanaceous d) Root knot nematode

168. Herbicide should not be applied in onion at:

a) Emergence-post crook stage b) Emergence-loop stage

c) Emergence-crook stage d) Loop-post crook

169. Which of the following is a miticide (acaricide)?

a) DDT b) Carbofuran

c) Chlorpyrifos d) Dicophol

170. Which system is used for hybrid seed production in onion?
a) Self-incompatibility b) Genetic male sterility
c) Cytoplasmic male sterility d) All of these

171. Homeostasis operates in the resistance mechanism known as:
a) Active resistance b) Field resistance
c) Horizontal resistance d) Vertical resistance

172. Bud necrosis of watermelon is a viral disease transmitted by?
a) Aphids b) Thrips
c) Whitefly d) Jassids

173. One of the progenitors of cultivated okra is:
a) *Abelmoschus crinitus* b) *Abelmoschus angulosus*
c) *Abelmoschus caillei* d) *Abelmoschus tuberculatus*

174. The desirable acidity level in tomato should be:
a) 0.1% b) 0.2%
c) 0.3% d) 0.4%

175. Which of the following is a triploid seedless watermelon?
a) Arka Madhura b) Arka Muthu
c) Arka Akash d) Arka Aishwarya

176. 'White heart' is a physiological disorder of:
a) Carrot b) Beet root
c) Muskmelon d) Watermelon

177. Which fungicide is most effective for controlling powdery mildew in pea crop?
a) Captan b) Mancozeb
c) Karathane d) Thiram

178. Potato leaf marker is commonly associated with male sterility line of:
a) Tomato b) Brinjal
c) Chilli d) Bell pepper

179. Transverse cotyledon cracking is a major problem in:
a) Pea b) French bean
c) Cowpea d) Pumpkin

180. Which of the following species of tomato is tolerant to drought?
a) *Solanum pimpinellifolium* b) *Solanum peruvium*
c) *Solanum penelli* d) *Solanum cheesmani*

181. The breeding method used for self-pollinated crops is:

a) Introduction b) Pure line selection
c) Mass selection d) Clonal selection

182. The thickness of covering material poly house should be:

a) 100 micron b) 150 micron
c) 200 micron d) 300 micron

183. What do you mean about NGS?

a) Next Generation Sequencing b) Novel Gene Sequence
c) New Genomic Sequences d) Novel Genomic Software

184. The UPOV (International Union for the Protection of New Plant varieties) headquarters is located is:

a) Paris b) Rome
c) Berne d) Geneva

185. Which of the following breeding methods is not suitable for asexually producing crops?

a) Clonal selection b) Pedigree breeding
c) Mass selection d) Mutation breeding

186. For construction of greenhouses, which of the following cladding material has the highest light transmission (95%) as well as long life?

a) Glass and polyethylene b) Polyvinyl chloride film
c) Polyvinyl fluoride film d) Ethylene tetrafluroethylene

187. The ratio between marketable crop yield and water used in evapotranspi ration is known as:

a) Water use efficiency b) Consumptive use efficiency
c) Economic irrigation efficiency d) Field water use efficiency

188. The variation due to uncontrolled factors is spoken of as:

a) Standard error b) Experimental error
c) Treatment effect d) All of these

189. Which aromatic plant is used as mosquito repellent?

a) Hops b) Citronella
c) Palmarosa d) Celery

190. Curcumin is extracted from:

a) Cucumber b) Capsicum
c) Ginger d) Turmeric

## Answers Key

| | | | | | | | | | | | | | |
|---|---|---|---|---|---|---|---|---|---|---|---|---|---|
| 1. | (a) | 2. | (c) | 3. | (b) | 4. | (b) | 5. | (a) | 6. | (d) | 7. | (b) |
| 8. | (a) | 9. | (d) | 10. | (b) | 11. | (b) | 12. | (b) | 13. | (c) | 14. | (c) |
| 15. | (b) | 16. | (a) | 17. | (c) | 18. | (a) | 19. | (b) | 20. | (a) | 21. | (c) |
| 22. | (c) | 23. | (c) | 24. | (b) | 25. | ) | 26. | ) | 27. | ) | 28. | ) |
| 29. | ) | 30. | ) | 31. | ) | 32. | ) | 33. | ) | 34. | ) | 35. | (b) |
| 36. | (a) | 37. | (c) | 38. | (a) | 39. | (b) | 40. | (a) | 41. | (c) | 42. | (b) |
| 43. | (c) | 44. | (c) | 45. | (b) | 46. | (b) | 47. | (d) | 48. | (a) | 49. | (c) |
| 50. | (c) | 51. | (d) | 52. | (b) | 53. | (a) | 54. | (c) | 55. | (a) | 56. | (d) |
| 57. | (d) | 58. | (c) | 59. | (c) | 60. | (c) | 61. | (a) | 62. | (c) | 63. | (a) |
| 64. | (c) | 65. | (c) | 66. | (b) | 67. | (b) | 68. | (a) | 69. | (a) | 70. | (a) |
| 71. | (a) | 72. | (b) | 73. | (a) | 74. | (c) | 75. | (c) | 76. | (d) | 77. | (a) |
| 78. | (c) | 79. | (b) | 80. | (d) | 81. | (a) | 82. | (c) | 83. | (a) | 84. | (a) |
| 85. | (c) | 86. | (c) | 87. | (a) | 88. | (a) | 89. | (c) | 90. | (c) | 91. | (c) |
| 92. | (a) | 93. | (c) | 94. | (a) | 95. | (b) | 96. | (b) | 97. | (a) | 98. | (b) |
| 99. | (c) | 100. | (d) | 101. | (a) | 102. | (c) | 103. | (b) | 104. | (b) | 105. | (a) |
| 106. | (b) | 107. | (c) | 108. | (d) | 109. | (a) | 110. | (a) | 111. | (c) | 112. | (a) |
| 113. | (a) | 114. | (c) | 115. | (a) | 116. | (a) | 117. | (d) | 118. | (d) | 119. | (d) |
| 120. | (c) | 121. | (d) | 122. | (b) | 123. | (c) | 124. | (d) | 125. | (a) | 126. | (d) |
| 127. | (d) | 128. | (b) | 129. | (c) | 130. | (c) | 131. | (a) | 132. | (a) | 133. | (b) |
| 134. | (c) | 135. | (a) | 136. | (b) | 137. | (a) | 138. | (b) | 139. | (b) | 140. | (b) |
| 141. | (c) | 142. | (a) | 143. | (a) | 144. | (a) | 145. | (b) | 146. | (d) | 147. | (a) |
| 148. | (c) | 149. | (b) | 150. | (a) | 151. | (a) | 152. | (a) | 153. | (d) | 154. | (d) |
| 155. | (b) | 156. | (b) | 157. | (d) | 158. | (b) | 159. | (b) | 160. | (b) | 161. | (c) |
| 162. | (d) | 163. | (d) | 164. | (c) | 165. | (b) | 166. | (b) | 167. | (a) | 168. | (a) |
| 169. | (d) | 170. | (c) | 171. | (c) | 172. | (b) | 173. | (d) | 174. | (d) | 175. | (a) |
| 176. | (d) | 177. | (c) | 178. | (a) | 179. | (b) | 180. | (d) | 181. | (b) | 182. | (c) |
| 183. | (a) | 184. | (d) | 185. | (c) | 186. | (a) | 187. | (d) | 188. | (b) | 189. | (b) |
| 190. | (d) | | | | | | | | | | | | |